The Structure of Time

International Library of Philosophy

Editor: Ted Honderich

A catalogue of books already published in the
International Library of Philosophy
will be found at the end of this volume

The Structure of Time

W. H. Newton-Smith

Balliol College, Oxford

ROUTLEDGE & KEGAN PAUL

London, Boston and Henley

First published in 1980
by Routledge & Kegan Paul Ltd
39 Store Street, London WC1E 7DD,
Broadway House, Newtown Road,
Henley-on-Thames, Oxon RG9 1EN and
9 Park Street, Boston, Mass. 02108, USA
Set in IBM Press Roman by
Hope Services, Abingdon
and printed in Great Britain by
St. Edmundsbury Press
© W. H. Newton-Smith, 1980

British Library Cataloguing in Publication Data

Newton-Smith, W H
The structure of time. – (International library
of philosophy).
1. Time
I. Title   II. Series
115   BD638   79–41330

ISBN 0 7100 0362 5

For my parents

CONTENTS

Contents

Contents

Now pray tell me what Time is? You know the very trite Saying of *St. Augustin, If no one asks me, I know; but if any Person should require me to tell him, I cannot*. But because *Mathematicians* frequently make use of Time, they ought to have a distinct Idea of the meaning of that Word, otherwise they are Quacks. My Auditors may therefore, on this Occasion, very justly require an Answer from me, which I shall now give, and that in the plainest and least ambiguous Expressions, avoiding as much as possible all trifling and empty words.

I. Barrow, *Lectiones Geometricae*, quoted from translation given by Čapek, 1975, p. 203

PREFACE

In 1715 Leibniz wrote to Caroline, Princess of Wales, to lament the decay of natural religion in England, a decay which he attributed in part to certain doctrines of Newton. The Princess passed the letters to the Reverend Samuel Clarke who took up the challenge on behalf of Newton. The ensuing exchange of correspondence between Leibniz and Clarke focused primarily on the conflict between Newton's absolutist conceptions of space and time and Leibniz's relationalist conceptions. While the theological concerns that motivated the exchange have not stood the test of time and while academic journals and books have displaced princesses as the medium of exchange, the controversy remains as lively today as it was in 1715. This book is intended as a contribution to this debate and to the issues concerning time which arise from it. In the course of this work it will be seen that neither the Leibnizian nor the Newtonian conceptions of time are tenable. In place of these I develop a pair of rival theories of time differing as starkly as those of Leibniz and Newton. In arriving at an articulation of this new and as yet unresolved controversy it will be argued that contrary to a venerable view still fashionable in some quarters, the investigation of the structure of time is, broadly speaking, an empirical matter. Such an investigation is not, however, an unproblematic and straightforward matter. For time's tenuous character generates cases of what Quine has called the underdetermination of theory by data.

An attempt has been made to present material available in the existing literature and to develop new material in a fashion which can be read by undergraduates without undue difficulty and by specialists without undue boredom. Explications of the necessary technical terms

xi

are given in the text and collected together for convenience in an appendix. The occasional more technical sections can be skipped by the nonspecialist without losing the main thread of the argument. With the exception of the Special Theory of Relativity, of which an introductory exposition is given, attention has not been given to current physical theories. While the conclusions reached are, I would argue, compatible with a correct construal of the relevant results in contemporary physics, this is a lacuna which I plan to make good in a later work. For, as Leibniz and Newton clearly realized, it would be rash to assume that the philosophical and the physical investigation of time are unrelated activities.

In developing the views presented I have benefited from discussions with many colleagues and students. Work with the late A. N. Prior on non-standard tense logics originally prompted me to explore the question of how one should choose between alternative hypotheses about the structure of time, so I am particularly grateful for the stimulus he provided. Discussions with J. R. Lucas awoke me from complacent relativistic slumbers and discussions with Professor P. Sandars assisted my understanding of matters relativistic. Their assistance is much appreciated. The causal efficacy of discussions with them and others should not be construed as betokening responsibility for the sins of omission or of commission in this work. In the production of the typescript I am thankful to Mrs Mary Bugge and my secretary, Miss Jane Long, for their efficiency and patience.

Novelists, more than others, appreciate how production depends on the personal ambience around the producer and for the ambience that made this work possible I am grateful to one of them, Dorris Heffron, my wife.

I

THE NATURE OF TIME

For what is time? Who can easily and briefly explain it? Who can even comprehend it in thought or put the answer into words? Yet is it not true that in conversation we refer to nothing more familiarly or knowingly than time? And surely we understand it when we speak of it; we understand it also when we hear another speak of it.

What, then, is time? If no one asks me, I know what it is. If I wish to explain it to him who asks me, I do not know.
Augustine, 1948, ch. XIV

1 WHAT IS TIME?

There is perhaps no concept which we use more extensively and more happily in our non-philosophical moments than that of time. Virtually nothing is ever said or thought save with the use of this concept. This is not just because we pay at least lip-service to time in the regular use of tenses but also because a whole host of important concepts have obvious temporal components. Not only is this concept used extensively, it is also used with facility. Children have no particular problem in acquiring a usable notion of time. And in everyday linguistic converse no one with the minimal normal linguistic competencies fails to grasp the meaning of utterances which make implicit or explicit use of the concept of time. However, we need only ask in a philosophical moment what it is we are talking about when we talk about time or the temporal aspects of things for the topic of our discourse to seem confused, complex and perplexing beyond comprehension. Often in desperation we make recourse to images, pictures, and metaphors in an attempt to make time comprehensible. Unfortunately these pictures usually have the effect of generating more mental cramps than they ease.

If we turn away from the host of pictures of time prevalent in any culture to the writings of philosophers, we find a bewildering maze of conflicting and often bizarre doctrines. The British neo-Hegelians hold time to be unreal on the grounds that the very concept of time is inherently incoherent. The Cambridge neo-Platonists tell us that time, being an attribute of God, is among the few really real things. Some tell us there is only, *sub specie eternitatis*, a sort of temporal arrangement of events; the past, present and future being mere anthropomorphic projections of the human mind. According to others, it is only the present that is really real. Some say that the advances of twentieth-century physics establish time to be really like space. Others tell us that physical theories give us no recourse but to temporalize space. And so on. . . .

The seemingly intractable difficulties involved in giving a true and enlightening answer to the question 'what is time?' reduced Augustine to praying for enlightenment. Augustine gave a lucid and oft-quoted statement of our predicament: 'What, then, is time? If no one asks me, I know: if I wish to explain it to one that asketh, I know not. . . .'[1] Others have been led to regard our question as not intractable but improper. Waismann takes it to be misguided on our part to seek a general account of time.

> whoever is able to understand the word 'time' in the various examples and to apply it, knows just 'what time is' and no formulation can give him a better understanding of it. The question: 'What is time?' leads us astray, since it causes us to seek an answer of the form 'Time is . . .', and there is no such answer.[2]

It seems, according to Waismann and Bouwsma,[3] that one ought to wallow in the nitty-gritty of our everyday temporal talk, hoping thereby to purge ourselves of the sickness of the mind whose symptom is the questing for illegitimate generalities. Before embracing either of these unpalatable alternatives — either that the task is simply beyond us or that there is no legitimate task at all — it is instructive to reflect briefly on the sources of time's intractability.

2 THE INTRACTABILITY OF TIME

Without meaning to be terribly enlightening we might truthfully say that time is *a system of temporal items* where by temporal items we understand things like instants, moments, durations and so on. This is unenlightening since, even granting its truth, we are accounting for the obscure notion of time in terms of the equally if not more obscure notions of instants, moments, durations. With regard to these temporal items most of us are, I suspect, rather like Augustine. In non-philosophical moments all is clear and the permeation of our thought and conversation with reference to temporal items is no barrier to comprehension. However, we need only ask ourselves in an idle moment just what instants are for all to seem totally obscure. For in general any type of item of which we can ostensively give instances seems less problematic than any type of abstract item about which we talk without being able to display examples. Temporal items, in virtue of being abstract items, seem more problematic than items like events, processes, occurrences and so forth which we think of as being in time. For we can have experience of items of this latter sort. If, Hume-like, we call up the impression of a particular moment, we at best produce the faded image of some event that happened at the moment. Like the Humean soul the moment itself entirely eludes our grasp. Time, then, in virtue of being a system of abstract temporal items, is a rich source of philosophical perplexity. Consequently, we expect a satisfactory general account of time to include an account of what we are talking about when we talk about these temporal items.

In answer to a question of the form 'what is X?' it is sometimes appropriate to point to examples or instances of X. In other cases it is appropriate to offer a definition or verbal equivalent of 'X'. But all verbal explanations, from the Oxford Dictionary's explanation of time as 'duration, continued existence' to Aristotle's explanation of time as 'the number of motion with respect to earlier and later' strike us as distinctly unsatisfactory. For even granting their truth their circularity offends. This is perhaps the crux of our difficulties. For time is not just an *abstract* beast but also it is a most *promiscuous* beast who regularly couples with equally elusive partners. *Prima facie* there are links between the concept of time and a host of other concepts including the following: motion, space, causality, change, entropy, human action, consciousness. Each of these concepts is in need of philosophical treatment and each requires reference to time in its elucidation. Thus, enlightening as it may be to explore the links between the concept of time and these other concepts, we are unlikely to arrive at some unique, non-circular, meaning-preserving analysis of the concept of time in terms of some other concept or complex of concepts. Time is just too basic to our entire conceptual framework to be captured in this way.

Philosophers interested in the question 'what is X?' (for some specific value of X) often begin by reflecting on the structure of ordinary everyday discourse about X. Such a technique is a fruitful starting point but it is usually no more than that. For in most cases there will be a range of substantive and important questions that can only be articulated once this is done. For instance, in the case of time it will be seen that we face deep and difficult metaphysical issues concerning the existence of facts (cf. ch. X, sect. 7). In addition, we wish to know not only how we do think of time (as revealed in ordinary discourse) but how we ought to think of time in light of contemporary physical theories. This means that any serious philosophical study of time must include discussions of the theories of Quantum Mechanics, Special and General Relativity (among others). This relevance of physical theory to philosophical reflections about time will be expressed by the description of time as a *theoretical* concept. Within this present book which is intended as a prolegomenon to a further study which will discuss time in Quantum Mechanics and General Relativity, space allows only for a treatment of the Special Theory of Relativity. My intention is to pave the way for this study by establishing that, contrary to a long and venerable philosophical tradition, the investigation of the structure of time is an empirical matter and as such cannot be conducted in an entirely *a priori* manner.

The problem of time's abstractness, promiscuity and theoretical character, is compounded by our own 'picture-making' inclinations. Faced with something so important to us as time, and faced with the difficulty of saying anything both intelligible and enlightening about time, we have a natural tendency to take refuge in analogies and pictures. Once adopted these pictures exert a powerful force and often not only fail to bring enlightenment but also compound our puzzlement. For instance, we are all familiar with the picture of time as a river. We think of ourselves as immobile on the bank while the events of life rush by. This picture suggests the possibility of rushing down-stream and experiencing some past event a second time and tampering with it a little. But we know we cannot in fact visit the past and *prima facie* at least there seems to be an incoherence in the supposition that we could. So time ceases to be just an ordinary river and becomes a most mysterious

4

river. Our puzzlement is thereby compounded. For the picture suggests possibilities of which we find it difficult to make sense. Consequently we are often faced with the negative task of thinking ourselves out of certain tempting pictures that cloud our understanding in our quest for enlightenment.

Such is the problem we face. In what follows I will grapple with the task of saying something both general and enlightening about some facets of time. The scope of the work is sweeping and a large number of problems are tackled, all of which relate to questions concerning the metric and the topology of time. Ideally one would like to embed the answers to particular questions about time within a general theory of time. In fact what will emerge is not one general theory of time but a pair of rival theories of time. Within the confines of this present work we will have to be satisfied by this *embarras de richesse*. For it will be shown that the choice between these theories must rest on resolving a major controversy within the philosophy of language. Neither of these theories has had particular currency in the history of the science and philosophy of time, and in the course of reaching this conclusion those theories that have had greatest historical currency will be seen to be untenable. A consideration of these theories, usually referred to as the relational and absolute theories of time, will provide a fruitful starting point. As these labels are often used to cover a rag-bag of differing claims about time I will employ the terms *Reductionism* and *Platonism* whose precise content will be given later in this chapter in terms of some subsidiary notions to whose explication I now turn.

3 THINGS IN TIME AND TEMPORAL ITEMS

The term 'things in time' will be used as a label to cover all types of item that are both in time and involve change. Thus, among the things in time are events, changes, processes, occurrences, happenings, incidents and so on. Clearly these notions themselves require conceptual elucidation. However, this task can be set aside here as our aim is only to sketch broad pictures of two rival theories of time. The set of all things in time that have happened, are happening or will happen will be called the *history set*. We are interested not only in what happens but also in the order in which things do happen. For describing this order we have a rich vocabulary which includes terms for various temporal relations such as being before or after, being earlier or later than, being simultaneous with and so on. We also employ temporal connectives such as 'while', 'and then', 'and next' etc. In addition we have a rich system of tenses which can be used to ascribe order to things in time. Lying behind all of these modes of expression is the idea that there is some objective ordering of things in time. I will use the term *ordered history system* to refer to the ordering of the history set.

Prima facie there is not just the ordered history system but also the times (moments, instants, intervals, etc.) at which the things in time occur. By *temporal items* I will mean things of this ilk which are ordered in a certain manner. By the *time system* I mean the ordering of the set of all temporal items. The question of the relation between the time system and the history system will be one of our primary concerns. The nature of the issue involved can be elucidated by considering one pair of the rival, historically important, and still prevalent theories of time mentioned above.

4 REDUCTIONISM

Leibniz, in his claim that time is nothing but the order of succession of creatures,[4] indicated a possible programme which if it could be successfully carried out would give a particularly satisfying answer to our question. This is a reductionist[5] programme which seeks to display the time system as a logical construction built up out of the history system. It is claimed that all talk about temporal items and the time system is merely a *façon de parler* which can be parsed in a meaning-equivalent way into talk about things in time. In developing his case the reductionist places great stress on the fact that whenever we have occasion to talk about some particular temporal item we identify it by citing some event which in fact occurred at that temporal item. For example, I might single out some moment as the moment I last turned a page of this book. Of course, we cannot equate the moment with that event. For we want to say that that moment is, say, the same moment as the moment you dropped your pen. As the same moment may have different identifying events the moment cannot be some one of these events. However, we can form the collection of all events simultaneous with any particular event used in identifying the moment. This collection, the reductionist claims, just is the moment.[6] To say that such and such an event occurred at such and such a moment is just to say that the event in question is a member of the set of events constituting the moment. To justify fully a programme devevloped from these beginnings the reductionist has to show that *all* talk apparently about temporal items can be treated as talk about things in time.

Clearly this is a most attractive project. For if it can be carried through it shows we need not think of temporal items as being ethereal, insubstantial, and unduly abstract. It should be noted that the reductionist ought not to characterize his position as that of denying the existence of temporal items. He ought rather to say that temporal items really do exist and claim to have given a particularly satisfying account of what these items are. Further, it shows us how to investigate the properties of time. For example, to settle the issue as to whether or not there was a first moment of time we need only settle the issue as to whether or not there was a first event. If we are thinking of time as a system of temporal items not reducible to history we are apt to be at a loss to see how we can answer this and other questions about the topology of the time system. Indeed, so attractive is the reductionist programme that many physicists and scientifically minded philosophers have come to believe that some form of this theory *must* be true, with the result that little attention has been given to fully and adequately articulating the theory, let alone to defending it.

5 PLATONISM

In sharp contrast to this view there is a view of the relation between the time system and the history system which I have called Platonism after the Cambridge neo-Platonists who were important articulators of the view. Such a view was expressed as follows by Barrow in his *Lectiones Geometricae*:

Common Sense, therefore allows Time to partake of Quantity, as the Measure of the Continuance of Things in their Being. But perhaps you may ask, whether Time was not before the World was created? And if Time does not flow in the Extramundane Space, where nothing is: A mere Vacuum? I answer, that since there was Space before the World was created, and that there now is an Extramundane, infinite Space (where God is present;) inasmuch as there might have been of old, and now may be, such, and so many Bodies, which then were not, and now are not; consequently Time existed before the World began, and does exist together with the World in the Extramundane Space, because 'tis possible that some Thing might have existed long before the World was made; and there may now be something in the Extramundane Space, capable of such a Continuance: Some *Sun* might have given Light long before; and at present this, or some other like it, may diffuse Light thro' Imaginary Spaces. Time therefore does not imply an actual Existence, but only the Capacity or Possibility of the Continuance of Existence; just as Space expresses the Capacity of a Magnitude contain'd in it. But you may perhaps wonder why I explain Time without Motion, and will say, does not Time imply Motion? I answer no, as to its absolute and intrinsic Nature; any more than it does Rest. The Quantity of Time, in itself, depends not on either of them; for whether Things move on, stand still; whether we sleep or wake, Time flows perpetually with an equal Tenor. If you suppose all the fixed Stars to have stood still from their Beginning; not the least portion of Time wou'd be lost by this; for so long as that Rest continues, so long has this Motion flowed. There may be what we may call *first* and *last*, beginning and ending together, (with Regard to the first Appearance and Disappearance of Things) even in that State of Tranquillity, which some Mind more perfect than ours may possibly comprehend. But as Magnitudes themselves are absolute *Quantums* Independent on all Kinds of Measure, tho' indeed we cannot tell what their Quantity is, unless we measure them; so Time is likewise a *Quantum* in itself, tho' in Order to find the Quantity of it, we are obliged to call in Motion to our Assistance, as a Measure whereby we may esteem and compare the Quantity of it; and thus Time as measurable signifies Motion; for if all Things were to continue at Rest, it would be impossible to find out by any Method whatsoever how much Time has elaps'd; and the several Ages wou'd roll on imperceptibly and undistinguish'd.[7]

We therefore shall always express Time by a right Line; first, indeed, taken or laid down at Pleasure, but whose Parts will exactly answer to the proportionable Parts of Time, as its Points do to the respective Instants of Time, and will aptly serve to represent them. Thus much for Time.[8]

It is not my intention to do historical justice to the Cambridge neo-Platonists. I intend only to capture a historically important and still prevalent view of time suggested to some by their writings. For a contemporary version of this view of time can be obtained in Swinburne.

While Swinburne does not explicitly espouse this view, I would argue that the theses he does advance commit him to a form of Platonism.[9] In this tradition time is thought to be a system of temporal items having real existence independently of the existence of the history system. Time is that in which world history is situated. To put the point in theological terms — if it had not pleased the Creator to produce world history the time system would none the less have existed. On this theory with its stress on the real existence of temporal items and their independence from things in time, the properties of the time system cannot depend on the properties of the history system. Time is taken to be continuous, linear, non-ending and non-beginning and is thought to possess these properties as a matter of necessity. While historically this view was generally developed and defended within a theological context, it will be seen that it can be and is defended outside such a context. For instance, one is committed to something like this view if one holds both that it is a necessary truth that there was no first moment of time and that it is a contingent matter whether there was a first event.[10]

In what follows it will be taken that the theses given below characterize respectively the reductionist and Platonist positions. The notion of a topological property of time is explicated in chapter III. Examples of topological properties include among others the following: having a beginning, having no beginning, being linear, having the same structure as a circle.

REDUCTIONISM

The Ontological Thesis of Reductionism

1 All assertions involving reference to time or to temporal items can be analysed in terms of assertions not involving such reference but involving instead reference only to things in time and to the temporal relations between things in time.

The Topological Thesis of Reductionism

2 Propositions ascribing topological properties to time are contingently true or false as the case may be.

PLATONISM

1 The Ontological Thesis of Platonism

The existence of temporal items is ontologically independent of the existence of things in time. Temporal relations between things in

9

time hold in virtue of temporal relations holding between the times at which the things in time occur.

The Topological Thesis of Platonism

2 Propositions ascribing topological properties to time are, if true, necessarily true.

While the pairs of theses given above go naturally together, they are logically independent. On the one hand, for instance, one might hold the second reductionist thesis without the first. In fact it will be argued that the first is untenable and the second is basically correct. That is, one might hold that we are committed to the existence of abstract temporal items which cannot be treated in a reductionist manner while maintaining that the character of the system of such items is to be empirically investigated. On the other hand one might hold, no doubt with less plausibility, the first reductionist thesis while denying the second. In this case one would treat temporal items reductionistically while holding that there are necessary truths about the temporal ordering of the set of events that entail ascribing a particular topological structure to time. For instance, someone might believe, in virtue of believing a particular form of the causal principle to be necessarily true, that it is not logically possible that there was a first event. However, we will see that the reductionist thesis does commit one to regarding the postulation of certain topological properties as empirical.[11]

Turning to the pair of theses which characterizes Platonism, one could hold that time possesses its topological properties as a matter of necessity and deny that temporal items exist independently of things in time. On the other hand, it is hard to see why one would want to hold that the system of temporal items exists independently of things in time without holding that time possesses its topological properties as a matter of necessity. For the ontological thesis entails that the system of temporal items would exist even if the history set was empty. In that case its properties could not depend on the character of the history system. However, in the framework of certain beliefs it would be plausible to hold the first but not the second Platonist thesis. For instance, suppose one believed that God in some sense produced a time system prior to producing world history and that he was free to produce one type of system rather than another. In this case one might not regard true assertions about the character of the time system as necessarily true. However, one would regard the existence of temporal items as ontologically independent of the existence of things in time.

6 A PLAN OF THE WORK

In the next chapter we will investigate the relation between time and change. It will be shown that the relation between time and change is not so intimate as to preclude the possibility of time without change. Once this is granted, reductionism in the form we have characterized it must be rejected. The topological structure of time will be the subject of discussion in the succeeding four chapters. The results arrived at will show that neither Platonism nor certain weaker versions of reductionism are tenable. In addition, it will emerge that in some regards theories concerning the topological structure of time can be underdetermined by all actual and possible data. That is, in some cases there can be logically incompatible theories concerning the topological structure of time where, no actual or possible observation would favour one theory over the other. Following this, in chapter VII, an account of the measurement of time will be developed. It will be seen that the phenomena of underdetermination may also arise in regard to the measurement of time. A version of the Special Theory of Relativity will be presented in chapter VIII with a view to showing how that theory can be consistently integrated into my accounts of the topology and metric of time. In chapter IX the question of the direction of time will be considered.

Platonism and reductionism are not the only general theories of time currently espoused, and in the final chapter (chapter X) a number of rival theories will be examined and found wanting. There are, however, two general theories of time that are compatible with the particular conclusions that will be reached in the course of the discussion of the topology and metric of time. These theories which can be best seen as embodying different reactions to the phenomenon of underdetermination will be characterized in the final chapter. As has been indicated earlier, the choice between these theories rests on issues whose resolution takes us outside the scope of the present work.

In very general terms we could say that the basically metaphysical theme pursued throughout this work concerns the relations between time and the world. At this level of generality, another theme deserving exploration could be characterized as the problem of the relation between time and consciousness. There appears to be an intimate link between time and consciousness, and without some exploration of this link the position reached in this essay might be held to involve both error and omission; error on the grounds that attention to certain features of consciousness would require revision in the claims about time advanced; and omission on the ground that so intimate and important is the link between time and consciousness that no general account of time could be of interest that did not explore this. One aspect of this general question which is of particular interest is the following. There is a long tradition which regards time or some aspects of time as being in some way mind-dependent, anthropomorphic, subjective, or psychological. In discussion of time these phrases have been used to cover a multitude of sins and are generally uncritically employed in a nebulous and virtually contentless way. Lying behind these characterizations is at least the following thesis. All (some) propositions about time or about the temporal aspects of things are such that they would be false in any world devoid of conscious beings. The thesis that this is so might be referred to as the thesis of the total (partial) *mind-dependence of time.* Unfortunately a consideration of this thesis cannot be undertaken within the confines of this work. I mention this thesis in order to partially define the areas which will be explored. In particular this thesis of the total or partial mind-dependence of time has been discussed in regard to the analysis of temporal indexicals and in regard to the problem of the status of 'temporal becoming'. Neither of these interesting topics can be explored within this work. While a fully satisfactory treatment of time must make reference to these issues I believe that some worthwhile progress to this end can be achieved even if these areas are set aside for the moment.[12]

II

TIME AND CHANGE

If it be a sufficient proof, that we have the idea of a vacuum, because we dispute and reason concerning it; we must for the same reason have the idea of time without any changeable existence; since there is no subject of dispute more frequent and common. But that we really have no such idea, is certain. For whence shou'd it be deriv'd? Point it out distinctly to us, that we may know its nature and qualities. But if you cannot point out *any such impression,* you may be certain you are mistaken, when you imagine you have *any such idea.* Hume, 1960, p. 65

1 REDUCTIONISM AND ARISTOTLE'S PRINCIPLE

In part, time's elusiveness arises from the fact that it is not an item given in experience. It is things in time, and neither temporal items nor time itself, of which we have experience. Objects, states of objects, changes in the states of objects, persons (ourselves and others), states of persons, changes in the states of persons, all of which are located in time, seem, because they can be experienced, less problematic than the ethereal instants, moments and durations of time itself. The reductionist seeks to capitalize on the *relatively* unproblematic character of things in time by accounting for time in terms of things in time — construing propositions about time and/or about the temporal aspects of things as modes of speaking about the sorts of things normally said to be in time. The viability of this admittedly attractive programme depends on the following crucial assumption: it is necessarily true that

13

there is no period of time without change taking place somewhere throughout that period. Hereafter I will refer to the claim that there is no period of time without change as *Aristotle's Principle*, or, for short, *AP*. While it will be necessary later to scrutinize critically the notion of change, for the moment we can understand 'change' simply as a generic term covering events, processes, episodes and so on.

It is easy to see both why the reductionist programme must presuppose *AP* to be a necessary truth and why this presupposition is *prima facie* plausible. For the reductionist must construe, roughly speaking, the notion of a period of time, say the period while I was typing this page, as the set of all events, processes and so forth that occurred simultaneously with my typing of the page.[1] Unless *AP* is a necessary truth, reductionism fails. For if it is not, there might be periods of empty time and there is *ex hypothesi* no possibility of treating talk about these empty periods of time as merely a manner of speaking about things occurring during those periods. This crucial assumption of the reductionist is *prima facie* plausible. For after all, we never experience the passage of time without having the experience of change.[2] When we are aware of time's passage, we are so aware in virtue of being aware of changes in our mental states.

Unless *AP* is a necessary truth the reductionist programme collapses. On the other hand, if *AP* is a necessary truth Platonism must be rejected. For the Platonist, time is a system of temporal items which necessarily exists independently of things in time. Given that the existence of a changing universe is a contingent matter, the Platonist is committed to the possibility of time without change. We shall see later (section 9 of this chapter) that the question of the status of *AP* is not of interest only in relation to its bearing on the Platonist–reductionist controversy. It has, for example, implications for the question of the difference between space and time.

2 CHANGES

In order to have a version of *AP* which is not trivially and uninterestingly true, it is necessary to delimit what is to count as constituting a change. We might begin by characterizing change with the following schema: something has such-and-such a property (or, lacks such-and-such a property) at one time and then later the same thing lacks that property (or, possesses that property). As it stands, the schema is in one way too restrictive and in another way not sufficiently restrictive. On the schema given, the occurrence of a change requires the persistence through time of some object that changes in respect of one or more of its properties. Certainly the changes we normally encounter involve persisting objects. But this need not be so. The box of matches on my table is not undergoing any apparent change at the moment. Suppose it were suddenly and instantaneously to cease to exist and in its place a tin of tobacco were to appear which had not previously existed. If this happened there would be a change, a most baffling and mysterious event would have occurred, though no persisting object would have changed. We can accommodate this case by modifying our schema as follows:

A change occurs *if and only if* some object has at some time some property and later that object lacks that property *or* some region of space is characterizable in such and such a way (i.e., as containing a match box) and later that same region of space is not so characterizable.

A change is thus constituted by an alteration in the properties of a persisting object *or* by an alteration of the properties of a region of space. As any change in an object involves change in a region of space, it might seem possible to characterize all changes in objects in terms of changes in spatial regions. However, in a sense the notion of change in a region of space presupposes the notion of change in objects. For we can only talk about changes in spatial regions if we can re-identify spatial regions through time and the criterion for sameness of region through time depends on the identity of some objects through time; namely, the objects which provide the frame of reference in relation to which we identify regions of space. Thus, to have a notion of change in spatial regions we must first have a notion of change in objects.

The schema as it stands is not restrictive enough. For instance, consider the proverbial death of Queen Anne. At one time it was present and at a later time it was not present. Are we then to say that this event has changed with regard to the property of being present? Some philosophers would not count this as a change at all. If we do count this as a change we will want to mark it off as a very special sort of change. For instance, very different implications follow from its being the case that the political situation in the country is undergoing a change from being stable to being unstable, than follow from its being the case that Queen Anne's death is now changing from being somewhat past to being still more past. In the former case it is sensible to suppose that we can do something to influence the changing or take steps to mitigate the effects of this change. It is not sensible to suppose either that there is something we can do about the change in Queen Anne's death or that this change has any implications of which we should be advised to be cognizant. What is important is the distinction between these cases and not the terms we use to draw it. I will say that anything satisfying the schema constitutes *a change* and mark the distinction by delimiting a class of changes to be called *non-trivial* changes. Any non-trivial change must be *expressible without the use of temporal indexicals.*[3] We can rule out as trivial the sort of change Queen Anne's death undergoes. Unless we take *AP* as asserting that during any interval of time there is *non-trivial* change it becomes trivially true that any period of time contains a dense flux of change. To see this let us suppose there could be some finite interval of time in the future during which no non-trivial change occurs. During that interval the event of the ending of all non-trivial change will undergo the trivial change of becoming more and more past.

We also have to exclude as trivial change what might be called *Goodman changes*. Suppose we introduce a predicate 'fred' defined as follows:

x is fred at *t* if and only if *x* is red at *t* and *t* is earlier than *t'* or *x* is not red at *t* and *t* is later than or the same time as *t'*.

Any object which is red before, at and after time *t'* will, at *t'*, change from being fred to no longer being fred. By defining predicates of this sort we can produce as many changes as we like during any period of time. Our previous restriction does not rule out as non-trivial change with respect to Goodman predicates. For Goodman change can be expressed without recourse to temporal indexicals. It seems we can rule out what we wish to rule out here by requiring that any case of non-trivial change must not be change with respect to a property which is such that its application at a time cannot be determined without knowledge of what the time is. An elaboration of this characterization of Goodman is given by Blackburn.[4] This manner of characterizing Goodman changes might be held to be objectionable on the grounds that it makes reference to how we recognize predicates as applying. It might be held that there could be beings who could recognize whether or not a Goodman predicate applied without first having to ascertain the time. However, the reductionist would not wish to defend *AP* on the grounds that any period of time contained at least Goodman changes. Even if the characterization given of Goodman change is inadequate the reductionist is likely to share our intuition that there is a genuine distinction between Goodman and non-Goodman changes.

3 INCONCEIVABILITY AND VERIFICATION

In spite of the fact that *AP* is widely taken to be necessarily true it has been supported by surprisingly little argument. Aside from an argument of Leibniz all standardly advanced arguments are of two related and unsatisfactory forms, which I shall call the inconceivability and verificationist arguments.

Leibniz argued in *New Essays Concerning Human Understanding* that

> if there were a vacuum in time, i.e. a duration without changes it would be impossible to determine its length. Whence, it comes that ... you cannot refute the one who would maintain that two worlds, the one of which succeeds the other, touch as to duration, so that the one necessarily begins when the other ends without the possibility of an interval.

Clearly it is not logically possible to measure the period of a temporal vacuum *directly* by, say, counting the swings of a pendulum or the ticks of a clock. However, we frequently appeal to theoretical considerations in assigning measures of duration to periods of time by the most indirect methods. For instance, such considerations are being invoked when one talks of the first three minutes of the Big Bang. I will suggest in my positive argument that the only contexts in which we might be led to hypothesize the existence of temporal vacua are contexts in which we would have to hypothesize vacua of determinate duration. In any event, if temporal vacua were not even indirectly measurable this would not establish any incoherence in the notion. For there is no ground for accepting Leibniz's implicit claim that time *must* be measurable. There might well be worlds in which periods of time filled with events could not be measured at all, either directly or indirectly. Things might be so chaotic that no self-congruent sequence of events could be identified to serve as the basis of a measurement system, in which case, as will be argued in section 9 of chapter VII, time would be unmeasurable.

A further claim that Leibniz is advancing can be brought out as

17

follows. Suppose S is a description of the universe which is *event-free* in the sense that it attributes properties and relations to the objects of the world without implying the existence of any change. Suppose we have such a description characterizing the state of the universe at an instant of time. Leibniz is asking what possible grounds we could have for saying that some such description characterized the world not at an instant but during a changeless interval. To put the point in other words, if S is a description of the universe during a putative temporal vacuum, what grounds could we have for saying that S obtains for an interval (i.e. for a multiplicity of instants) as opposed to a single instant? This objection of Leibniz will be met later by providing a description of a context in which it would be more reasonable to take some such description as holding for a changeless interval rather than for an instant.

On the inconceivability argument empty time is held to be logically impossible on the grounds that it is inconceivable. If we take 'inconceivable' as meaning logically inconceivable this is not an argument for the claim that empty time is logically impossible but a mere re-statement of that claim. In a weaker sense of 'inconceivable' empty time is uncontentiously inconceivable but its being inconceivable in this weaker sense does not entail its being logically impossible. We might understand the claim that some state of affairs, p, is inconceivable as the claim that we cannot imagine what it would be like to experience that state of affairs. In this sense empty time is inconceivable. For as we noted previously, it is incoherent to talk of someone's experiencing empty time. If one is aware of experiencing at all there will at least be changes in one's mental states of which one is aware in virtue of which one knows time to have passed. As it is incoherent to describe some subject of experience as experiencing the passage of time in the absence of *all change*, we cannot imagine what it would be like to be in this state.

In this sense of inconceivability, the inconceivability of some state of affairs, p, does not entail that p is not logically possible. To borrow an example of Shoemaker's[5] — we cannot imagine what it would be like to experience a world in which the preconditions for experience were not satisfied. But for all that, there may be such a world. We could put the point with regard to time this way. There is no sensible use for the sentence 'there is now absolutely no change occurring anywhere'. It remains possible that we can describe a sensible use for the past- or future-tensed versions of that sentence.

It has been suggested[6] that we are erroneously seduced by certain pictures which we are prone to use when thinking of time into thinking that the notion of empty time is not empty. We might, say, be thinking of events as related to time as coloured beads on a wire are related to the wire. We may conjecture — 'What could be easier than empty time? It is just like a stretch of wire without beads.' These pictures are misleading and may prompt us to take as possible what is in fact incoherent. But even if it is these unsatisfactory pictures that incline us to toy with the notion of empty time, it may none the less be the case that the notion is not without content. At the same time it is possible that we are just misled by these pictures, and thus the onus is on one who thinks it makes sense to talk of empty time to display the content of that notion. One way of doing this is to describe circumstances in which it would be reasonable to say there had been or was going to be a temporal vacuum. This will be done following a brief discussion of the verificationist argument.

If we assume a proposition to be meaningless if it does not admit of verification even in principle, and if we construe what counts as verification in a sufficiently tough-minded way, we will be committed to regarding any hypothesis asserting the occurrence of changeless time as meaningless. For instance, if we admit as verificiation only verification on the basis of direct observation by the senses, then, for the reasons already given, the hypothesis of empty time does not admit of verification. But construing verification in this way gives us a principle of significance which is entirely implausible. For it would require us to regard a wide range of theoretical assertions as being devoid of meaning. I shall argue that if the notion of verification is construed sufficiently liberally to provide a principle of significance which is not entirely implausible, then we could have evidence warranting us in asserting the existence of temporal vacua.

4 TOWARDS CHANGELESS TIME

Sydney Shoemaker has recently advanced an argument for the possibility of empty time.[7] His strategy is to describe a logically possible world and to argue that the best account the inhabitants of such a world could make of their world is one that commits them to positing the occurrence of changeless periods of time. Shoemaker's argument, while open to objections, is plausible and can be amplified to make it considerably more plausible. Rather than consider his case in detail, I will strengthen his conclusion by offering the same style of argument based on a different fantasy world.[8] I will also offer a second argument based not so much on considerations of an all-out fantasy world, but on considerations of a possible theory about this world.

Basically, both Shoemaker's argument and mine are designed to show that talk of time without change has sense through providing a description of conditions under which we would be warranted in asserting the existence of temporal vacua. This sort of argument may establish that talk of temporal vacua has sense; it does not, however, give us a full account of what that sense is. I shall turn to this further question in section 10 of this chapter. In sections 6 and 7 following the development of my argument, I shall articulate the reasonable grounds we have for being reluctant to admit the existence of changeless time. Our unwillingness to admit exceptions to *AP* stems from deep and important methodological principles relating time and causality.

For the sake of this argument imagine a world containing nothing but a rather small number of objects and persons which look rather like the familiar macroscopic objects and persons of this world. The objects and persons are such that from time to time each disappears to reappear some time later. As the objects which disappear cannot be found elsewhere in space we conclude that they have temporarily ceased to exist. As in the actual world, the changes in objects while existing appear to be generally continuous. While I, as an inhabitant of this world, would notice that the other objects and persons disappear, I would not of course notice that I disappear. However, I am inclined to accept the testimony of otherwise reliable observers that this happens to me as well because I notice that this happens to them and I realize that their claims would, if true, have explanatory force. For the occasions on which I am told that I have been out of existence for a period of time are occasions in which it appears to me that objects have, instantaneously and discontinuously, undergone a change of state. Between such states I would normally expect there to be a smooth transition through a sequence of intervening states taking a certain period of time which is equal to the duration of time during which it is claimed I have been out of existence. It should be noted that this fantasy world raises questions of the criterion of identity of persons and objects across temporal gaps. I would argue that there can be identity across temporal gaps in certain contexts. But in any event, the argument goes through if we prefer to say that the objects and persons in this world cease to exist and that numerically distinct qualitatively identical objects later come into existence.

Let us suppose further that there are apparent regularities in the disappearances, and having selected some object as a clock, we discover it to be possible to calculate for each object what I will call its *vanishing function*. The vanishing function tells us how long each object will exist before disappearing, and tell us for how long it will disappear. Unhappily our clock will from time to time itself disappear. However, we astutely select an object to serve as a clock that does not disappear very often and does not disappear for very long. When it does reappear we advance it by an amount determined by reference to the occurrences in its absence of types of events, occurrences of which we have previously timed. Thus we discover for each object, 0, that it disappears for Δt_0 time units every Δn_0 time units.

We also discover that there is some observable, variable property of objects correlated with disappearances in the following manner. The parameter in question takes on values in a certain range if and only if an object is about to disappear and the value of the parameter on these occasions is a function of the time for which the object subsequently disappears. As with the vanishing function this regularity is invariably corroborated by observation. Further, let us suppose that in some cases we can control the parameter associated with subsequent disappearances. That is, by manipulating the value of this parameter we are able to make objects disappear for a predetermined period of time. And we are able to bring it about that objects which would otherwise disappear at a certain time for a certain period either do not disappear or disappear for a different length of time. In addition let us suppose that we detect in the area vacated by a vanishing object or person a strong local unchanging force field that goes to zero simultaneously with the reappearance of the vanished object. The field extends somewhat beyond the boundary of the object with an intensity that makes it difficult to move another object into this general area and makes it impossible to move an object into the precise place vacated by the vanishing object. That is, the field operates as a sort of place holder for vanished objects.

Sometimes in this possible world there are just a few objects missing, on other occasions there are a lot of objects missing. An inventory of all objects is compiled and their vanishing functions are determined. By elaborate computation it is discovered that, say, in the year 1984 at the stroke of midnight each object is scheduled to have a period of disappearance. Just prior to that time everything does take on its characteristic prior-to-disappearance state. At the precise time in question we notice nothing special. Perhaps we merely raise our glasses and drink a toast to the time period that has slipped by with total ease. Those of us happily blessed with short disappearance times will be able to notice the other objects and persons with longer disappearance times come back into existence.

While the postulation of the occurrence of this period of time without change fits the observable facts, the observable facts do not force these postulations on us. We could avoid positing the temporal vacuum by positing more complicated vanishing functions that allow for an exception to the projected functions every so many years. We could similarly say that there are exceptions to the projected correlations between vanishings and the observed prior-to-vanishing state. That is, there certainly are more complicated projections that fit the observable facts equally which would avoid a commitment to a temporal vacuum.

It will not be open for one who prefers the more complicated projections to argue his case by appeal to the incoherence or vacuity of the posit of a temporal vacuum. For in the absence of an independent argument for the thesis that there is such an incoherence or vacuity, this claim is just what is at stake. It can be urged that the complexity and *ad hoc*ness of the projections designed to avoid a commitment to temporal vacua are a fair price to pay for the goal of not multiplying the unobservables of a theory (in this case the unobservable being the temporal vacuum) beyond what is necessary to fit the observable facts. But to make out our case we do not need to tell the story in such a way as to make one set of projections seem overwhelmingly preferable to the other. It is enough that we can imagine a sensible debate between the proponents of the rival projections. Neither is committed to asserting anything either vacuous or incoherent.

There is a major objection to the projections which involve positing temporal vacua. This is the following: in order to be justified in positing vacua we need to be justified in regarding the observed correlations as law-like and not merely accidental. It may be argued we are not so justified unless we can provide some causal explanation of the disappearances and the reappearances. The disappearances do not raise any particular conceptual problems. We might come to regard the parameter referred to earlier as the causal element. That is, taking on such a state causes an object to vanish and to be replaced by the static force field. However, in order to have a causal account of the reappearance we need to make assumptions about causality that are at the very least unusual. For in the story as told the force fields are un-changing (otherwise the period when everything vanishes will not be a period without change). Thus, it cannot be a change in the field that causes the reappearance of the object. Nor can we suppose, in the context of the argument, that some changes in some existing object or objects causes an object to reappear. For in the hypothetical totally empty period of time there are, *ex hypothesi*, no objects at all. Shoemaker, in considering this objection to his argument,[9] suggests that we could regard the following sort of causality as operative. Merely being in a certain state (without change) for a certain period of time is causally sufficient for some change to then take place. That is, a field's being in such and such a state for the period of its associated objects' disappearance is a causally sufficient condition for the production of its associated object. If this sort of causality is incoherent, no argument of the style I have outlined will work. For the positing of any temporal vacuum prompts the question, what causes the vacuum to end? And unless some possible answer can be sketched, the grounds for regarding the projections involving temporal vacua as law-like are seriously undermined. I will return to this objection in sections 6 and 7 of this chapter following a consideration of Shoemaker's argument and an additional argument of my own supporting the contention that the notion of a temporal vacuum is coherent and non-vacuous.

I give a very brief account of the fantasy world described by Shoemaker in order to indicate the weakness in this argument that is avoided in mine. Shoemaker imagines a world divided into three regions, A, B, and C. It seems to the inhabitants of region A that there is a total cessation of all change in the B region for one year every three years and a total cessation of all change in the C region for one year every four years. The inhabitants of the B and C regions report a total cessation of change in the A region for one year every five years. This basic situation is elaborated and complicated in various ways by Shoemaker. We are to suppose that the inhabitants are justified in projecting the observed regularities holding for each region. Hence they are justified in holding that there is a year every sixty years in which there is no change at all in any of the three regions. Shoemaker notes that in order to have a causal account of the ending of these periods of total changelessness, the inhabitants will have to suppose that this mere non-occurrence of change for a certain period of time is a causally sufficient condition for the beginning once again of change.

As Shoemaker tells his story, the inhabitants of each region have no experience of this strange sort of causality during periods of normal change. Against this background one might think it would be more reasonable for the inhabitants to retain their ordinary beliefs about causality and adopt some hypothesis not involving a commitment to a temporal vacuum. Noting this objection Shoemaker suggests in a footnote that we can suppose the inhabitants to discover other phenomena which seem to involve this sort of causality. However there is a problem with this. For if in periods of normal change there are some changes which occur merely as a result of the passage of time, these changes would still continue during the partial freezes. In this case there will be no grounds for positing the occurrence of a year devoid of all change once every sixty years. I have sought to meet this difficulty by supposing both that the inhabitants of my world are familiar with this sort of causality and that it operates in such a way as to provide grounds for positing periods of time devoid of all change.

5 QUANTIZED CHANGE

We can proceed to establish the possibility of a temporal vacuum without such an extreme fantasy by describing a particular system of beliefs we might have grounds for adopting about the actual world. These beliefs would be consistent with the character of our everyday experience of the world and would commit us to positing the existence of temporal vacua. The story will be much less dramatic than either Shoemaker's or mine. It will not give us the trouble-free Shoemakerian years but tiny little intervals of time devoid of change. But the smallest temporal vacuum is just as threatening to the reductionist as the really big one. My strategy will be to give the postulates of a type of theory we could have grounds for adopting which would involve positing empty time. While the ingredients of this sort of armchair theory have been toyed with by some physicists, it is well outside current paradigms and is, for reasons to be given later, the sort of theory physicists would be most reluctant to embrace. The basic tenets of this theory are as follows:

1 All observable change in the world is to be explained by reference to the properties and behaviour of some finite class of types of entities to be called particles, collections of which constitute the macroscopic objects of the world.

2 In so far as fields are involved in the theory, all fields have as sources particles or systems of particles and change only in response to changes in their sources. The point of this is to rule out the possibility that there could be change, change in the state of a field, in the absence of any change in the particles.

3 There are only a finite number of particles in the physical world. We assume a finite unbounded cosmological model of the world and assume the particles have some finite size.

4 All change that these particles undergo is *discrete quantized* change. That is, if some particle is in a state S it is in that state for some finite interval of time and there is some next state that it takes on. The change of state is instantaneous and the interval between these changes is much smaller than could ever be experimentally determined.

5 For any particle there is some minimal time interval between any two adjacent changes.

Assuming these postulates hold of our world it follows that during the last hour there were only a finite number of these micro-changes, and that there were lots of tiny little periods of changeless time. For example, what we would describe as the continuous motion of my arm as I wave it about, was in fact constituted by a finite sequence of perceptually indistinguishable jumps of the system of particles comprising my arm. Moving particles will be spatio-temporally discontinuous. If particle p is at spot x at time t and has as its next position x', and is there at t', it will jump instantaneously from x to x' without occupying any intervening positions. One might wonder whether one is licensed to talk of sameness of particle over these spatial jumps. But for present purposes it does not matter whether we think the particle at x', t' is the same as the particle at x, t or is a new particle standing in some causal relation to the formerly existing particle. Such a theory represents a radical departure from any theory currently in fashion. But there is no incoherence involved and if current theories came to face extreme difficulties, we might well adopt some theory of discrete change which fitted the observable facts at least as well as existing theories and which was more elegant and simple in its formulation.

The crucial tenet in this armchair theory is postulate 4. In some theories, some parameters have been taken as changing their values discretely. For instance, in the Bohr-Rutherford model of the atom, the parameter giving the orbit level of an electron changes in a discrete quantized manner. The electron is in some orbit and if it changes orbit level at all there is a next orbit level which it reaches by instantaneous jump. In this theory other parameters change continuously. For instance, the electron rotates around the nucleus in a continuous manner. What our theory posits is the treatment of *all* parameters as discrete quantized parameters. Theories of this form are at least *toyed* with by physicists.[10] But given the success of current theories that do not involve discrete quantized change there is little incentive to try this sort of alternative. Poincaré who, as we noted, argued at one stage for the necessity of continuous space, considered seriously the possibility of discrete quantized change. In a late essay on Planck he explored the consequences of the possibility that there are only a finite number of possible states of the universe.

The universe will jump abruptly from one state to another; but during the interval it will remain motionless and the different instants during which it remains in the same state will be indiscernible; thus, we would obtain discontinuous change of time, and *atoms of time*.[11]

Harré[12] has claimed, on the basis of an argument which derives, he says, from Boscovitch,[13] that no parameter can change its value discontinuously if time is continuous; that is, if time has the same structure as an interval of the real number line.[14] This argument, given below, would, if cogent, undermine my strategy and would be of greater interest even apart from its bearing on *AP*.

Suppose now that in some continuous sequence of times there should be discontinuous action. Let the action have measure *a* in the lower segment of the sequence of instants, and *b* in the upper segment, and the difference between *a* and *b* be always greater than zero. If the sequence of instants is genuinely continuous then there has to be an instant *Q*, the point of discontinuity of the action, which, being a point in a continuous sequence of instants, has to be able to be considered either as the upper limit of a lower segment, in which case the action has measure *b* at *Q*, or as the lower limit of the upper segment, and thus as the last member of the lower segment, in which the action has measure *a* at *Q*. So, either at *Q* the action has measures of both *a* and *b*, which is a contradiction, or, the action has no measure at *Q* which is contrary to hypothesis. In article 49 Boscovitch expresses the dilemma as follows: 'On the one hand there must be at any instant some state so that at no time can the thing be without some state of that kind, whilst on the other hand it can never have two (different) states of the same kind simultaneously.' Therefore there cannot be discontinuous action in continuous time.[15]

Suppose for the moment we take Harré's argument as cogent. In that case we can only adopt a discrete quantized model of change on the assumption that time is not continuous. I will argue (ch. V later) that we are free to treat time as either continuous or merely dense. Thus Harré's argument would at best force one who adopts a discrete quantized model of microscopic change to drop continuity in favour of mere density.

While the claim that there can be no discontinuous change of a parameter in continuous time does not thereby rule out the possibility that all change is ultimately quantized in the sense outlined, Harré's argument in favour of the claim is not cogent. We should accept that at any instance the parameter in question has some value (*a* or *b*), and that it does not have more than one value (not both *a* and *b*). Call these Boscovitch's constraints. Harré illegitimately introduces *Q* as designating *the instant of discontinuity*. If we introduce the term *Q* in this way, *Q* is not a uniquely identifying name. For *Q* designates either the last instant at which the parameter has value *a* or the first instant at which it has value *b*, but not both. So long as we regard *Q* one way or the other and not both ways at the same time we satisfy Boscovitch's constraints. Presumably Harré feels, given his 'realism', that, *Q* being an instant of time, the parameter either has value *b* or it does not have the value *b* and its being one or the other should not be a matter of stipulation on our part.[16] However, as introduced, the term *Q* does not designate a unique instant. We have to decide whether it designates the last instant at which the parameter has the value *a* or the first instance at which it has value *b*. Once we have so designated the referent *Q*, trivially, the value of the parameter at *Q* is not a matter of stipulation. In the former case *Q* is a greatest lower bound and in the latter case *Q* is a least upper bound.[17] So long as we introduce *Q* one way or the other, there is no problem. Thus Harré's argument does not rule out discontinuous change in continuous time.

In arguing for the logical possibility of empty time we have considered a fantasy world and a hypothetical theory. There is an actual theory, the Theory of General Relativity (hereafter cited as *GR*, which might be held to establish that possibility of empty time. The crux of

GR is the family of equations known as the field equations. By making assumptions concerning the value of various parameters in these equations we can solve them to obtain descriptions of possible spacetimes. Some of these solutions or models represent a universe expanding from a point (the Big Bang mode); others represent the universe as a homogeneous rotating sphere (the Gödel model). Among the plurality of further models are those corresponding to *vacuum solutions* of the field equations. A vacuum solution is obtained by assuming that the universe characterized by the model contains no matter or radiation whatsoever. Interestingly the term which describes the structure of the spacetime, the metric tensor, does not vanish in all vacuum solutions. If we take it that any universe compatible with *GR* is physically possible, this suggests that completely empty spacetimes are physically possible. For the metric tensor in vacuum solutions characterizes the structure of a spacetime devoid of all matter and events. Indeed, there is not a unique structure. For some empty universes have a flat spacetime and others have spacetimes which are curved.

As the *GR* is a respectable theory some have been inclined to regard this result as sounding the death knell of any reductionist account of space, time or spacetime. That is, it is taken that as this result in physics shows something to be possible which according to the reductionist is impossible, reductionism has to be rejected. Certainly this result ought to prompt us to examine critically the assumptions of the reductionist as we have been doing. However, it would be far too swift to conclude that *GR* has conclusively established the untenability of reductionism. For there is an alternative response to these results which is to argue that since a notion of a spacetime totally empty of contents is empty of content, some modification should be made in the field equations of *GR* to block these problematic solutions. And attempts have been made by physicists to produce rivals to *GR* which do not admit of vacuum solutions. Consequently we should conclude that, interesting as the vacuum solutions of *GR* are, their existence is not telling one way or the other in the reductionist–Platonist controversy, at least at this stage of scientific development.

6 DATE CAUSALITY

Admitting *any case* of a temporal vacuum will force us to modify certain well-entrenched beliefs about causality. In displaying this it will be fruitful first to articulate the principle about causality which would have to be abandoned and to explicate the grounds we have for holding that principle. The principle which we will call the principle of the *acausality of time* amounts, roughly, to the claim that time is not causally relevant to the occurrence of change in things.

Many explanations which we employ in scientific contexts fit the deductive nomological model of explanation.[18] We explain why some event occurred, say this substance's turning yellow, by citing features of the items involved in the event (the substance is sodium), by citing features of the history of the item involved (the sodium was placed in a flame just before turning yellow), and by citing certain regularities which we take to be law-like (other things being equal, any substance which is made of sodium and is placed in a flame turns yellow). That is, we explain occurrences of some type by discovering appropriate covering law-like regularities. In any actual case we cannot examine every factor involved. Consequently we make judgments concerning factors that are relevant and investigate them. If we fail to come up with the appropriate regularity, we may decide to investigate factors that we initially took to be irrelevant. In any event there is one factor which we never initially take to be relevant and which, no matter how poorly our quest for regularities fares, we do not attempt to bring in. This is *the time* of the occurrence of the event to be explained. For instance, suppose some substance, *S*, is introduced into a flame and turns green. We would never dream of thinking that it turned green because it was put into a flame at *just that time*. To think that the date was causally relevant in this case is to take quite literally the idea that *the time was ripe for change*. This view that the date of an occurrence is causally irrelevant to that occurrence is one ingredient in the principle of the acausality of time and will be referred to as the exclusion of *date causality*.

Apparent regularities that seemed to involve date causality could take one of two forms. Either the generalization might state that if such and such conditions, *C*, are satisfied at the *particular* time, *t*, then such and such will occur; or it might state that if such and such conditions, *C*, are satisfied at any time, *t*, having some property *F*, then such and such will occur at *t*. In the latter case the predicate, *F*, picks out some class of times. We can dismiss out of hand the first alternative which I will refer to as *particular date causality*. For if at some other time different from time *t* the conditions *C* are satisfied and the result is the same, we have shown that the conditions themselves are sufficient and the time at which they obtained is irrelevant. Or, if we find that at no other time does the result ensue if the conditions are satisfied, we have no grounds for regarding the occurrence of the result in the presence of the satisfaction of the conditions C at time t as anything more than a coincidence.

It is not so easy to dismiss the second alternative which I will call *general date causality*. Indeed, sometimes we *appear* to make use of it. For example, suppose there is a compass in the room which deflects at four o'clock. In answer to my quest for an explanation, you say that it happened because it was four o'clock. When I press you to explain what that had to do with the occurrence in question, you cite a generalization to the effect that every day at four o'clock a heavy lorry passes by and vibrates the house causing the needle to deflect. In this case citing the date as the causal factor is merely a manner of speaking. The causal factor is in fact not the time but some event that occurs regularly at that time each day: namely, the passage of the lorry.

It is clear that physicists simply do not countenance the possibility that the time of an occurrence is causally relevant to that occurrence. Indeed, Maxwell,[19] among others, has enshrined this prohibition in a general maxim:

> The difference between one event and another does not depend on the mere difference of the times or the places at which they occur, but only on the differences in the nature, configuration, or motion of the bodies concerned.[20]

One component in Maxwell's maxim is the principle of the *homogeneity of time* — time is the same everywhere and everywhen. That is, if time is homogeneous particular times are indistinguishable with regard to causal powers in virtue of having no causal powers. It is a deep and well-entrenched assumption of contemporary physics that time is homogeneous, i.e., that date causality never arises. That this is so is seen from the fact that physical laws are invariant under any linear transformation of the form $t' \rightarrow t + C$. It is not at all clear what status should be accorded to a maxim excluding date causality. According to the reductionist date causality is incoherent. For there is just nothing to an instant over and above those things obtaining at the instant. Therefore, an instant can be vested with causal powers only in the sense that something present at that instant has the causal power. One might think it a merit of the reductionist position that it explains why we never even toy with the idea that instants have causal powers. But in view of the difficulties in the reductionist thesis already noted and others to be noted, we cannot rest with this explanation.

Lucas[21] has argued for adopting the principle of the acausality of time as a *regulative principle* on the grounds that unless it is adopted we could discover no causal laws. For the discovery of causal laws presupposes the repeatability of experiments. As we cannot repeat experiments with *all* factors the same, we must have principles of irrelevance. Time must be included in the irrelevant factors for we cannot, so to speak, hold the time constant. This sort of consideration does, as we remarked above, seem to indicate that we would never be warranted in postulating the occurrence of particular date causality. On the other hand, general date causality seems immune to this style of pragmatic argument. For the condition, say four o'clock on Mondays, is repeatable.

In any event Lucas's argument would at best establish that as a matter of fact we could not begin to do sensible physics if we took time as being in these ways an important causal factor. It remains possible that we could begin by taking time as *not always all that relevant*, and having on that basis got some tentative theory, find it advantageous to posit the causal relevance of time. Suppose for instance that some event-type, E, has occurrences at all times, $t = t_0 + dn$ (where n ranges over the integers and d is our unit of time), if such and such other conditions were satisfied. Suppose further that if these conditions are satisfied at other times E does not occur. Might we not be forced to take seriously the hypothesis that the times given by the equation above were causally responsible for producing the particular events of type E?

If we found that if a certain condition C was satisfied at certain times, an event of type E occurred but that E did not occur at other times even if the condition C was satisfied, we would seek to discover some other conditions C_i, which obtained along with the condition C in cases where E occurred and was absent when condition C was satisfied at the other times. For instance, suppose an appropriate time was four o'clock on any day. Suppose our criterion for its being four o'clock is that the earth and sun are in a certain relation, a relation that does not obtain at any other time of the day. Hence we could seek to explain the occurrence of events of type E on the basis of the holding of condition C, and on the holding of such and such a spatial relationship between the earth and the sun. However, this attempt to avoid date causality would not work if the times having the power to produce E in condition C, were, say, four o'clock on Mondays of even years, five o'clock on Tuesdays of odd years. In this case there is no configuration of the solar system that is common to all and only those times with the power in question. Even in this situation we should be inclined to regard our generalization (If $C(t)$ and $t = t_0 + dn$, for some n then $E(t)$) *either* as an accidental generalization giving only the times at which E occurs in presence of C; *or* as a law-like generalization the acceptance of which is based on the assumption that those times such that $t = t_0 + dn$ have some other factor in common which is the causally relevant factor. There seem to be at least three sources of this reluctance. The first is the obvious *pragmatic* fact that we have had great success in finding causal explanations which do not involve date causality. The second factor is our *ontological qualms*. We should like if possible to minimize our ontology. Hence we should like to be reductionists (if only it worked). To allow date causality is not just to admit that instants are not collections of events and may exist independently of events, it is to make them very real individuals by giving them *causal powers*.

Thirdly, and most importantly, we do not see such causal explanations of the occurrence of events which appeal in effect to the ripeness of time as being *good explanations*. For suppose I explain the occurrence of E by citing the fulfilment of certain conditions C and by citing a law-like regularity connecting Cs and Es. In this context we are interested in discovering facts about Cs and Es and more basic regularities which account for the regular connection between Cs and Es. We seek to construct a system of generalizations in which this generalization is seen as a consequence of other more basic generalizations. To do this we have to be able to get hold of the Cs and Es and discover something further about them. Suppose now that our generalization involves putative date causality. It is hard to see what possible further generalization we could have arrived at that would account for the fact that some times and not other times have the power to produce E in the presence of C. For we cannot get hold of the times in question to discover further properties about them. If we discover some factor common to all and only those times, we would have ground for hypothesizing that factor to be the causally relevant one, in which case we do not have a putative case of date causality. If we do construct theories that involve the supposition of date causality we reach quickly the level of brute regularity that cannot be further accounted for.[22]

It seems then that we would never be warranted in positing the occurrence of particular date causality. However there might be contexts in which it would be warranted in positing the occurrence of general date causality. None the less, we would be extremely reluctant to do so and can articulate reasonable grounds for being reluctant to do so. And as we could always plead ignorance of the true causal explanations of cases that seem to call for the postulation of general date causality without being refuted by empirical observations, we are free to adopt a methodological principle excluding general date causality.

7 DURATION CAUSALITY

We do not take seriously the suggestion that the mere passage of time is causally relevant to anything. If some object changes its state after having been in that state for some period of time, Δt, we would not think that its having been in that state for that period was causally sufficient for it to change. To think otherwise would be to think that merely being in a certain state for some time could bring about the termination of that state. Certainly we might say that being an unrefrigerated egg for ten days is sufficient condition for becoming a bad egg. But this is because we suppose there to be some process going on in the unrefrigerated egg that takes this amount of time to bring about the change in question. However, we can tell stories of contexts in which it would be reasonable to attribute causal powers to the mere passage of time. Suppose we live in a world much like this except that billiard balls, if immersed in water for one hour, suddenly (instantaneously and discretely) change colour. We investigate and can find no causal influences impinging on the balls at or just before the time of a colour transformation. Even when isolated from all influences which we think might be causally relevant, this process is still observed. We investigate the structure of the balls themselves and as far as we can tell nothing is going on in the balls (or in the water) during the period. That is, by examining a ball you cannot tell how long it has been immersed. There are no properties which can be discovered which differentiate balls which have been immersed for almost an hour from those that have been immersed for a half hour; or, indeed, from balls that are not immersed. In such a situation we would be warranted in supposing that the passage of time had a causal influence. For it would be entirely *ad hoc* to insist that either there was some undetected causal influence impinging on the ball or that there was some undiscovered process going on in the ball.

Unlike the case considered earlier of particular date causality, repeatable experiments can be used in testing hypotheses involving the postulation of what I will call *duration causality*. For instance, we can come to have evidence that the mere passage of time has a causal influence on the balls by repeatedly immersing them while varying all other factors that might possibly be relevant. Duration causality has not been excluded from serious consideration because the null result was observed to obtain in experiments of this sort. Rather, it is excluded *ab initio* as we cannot see how the mere passage of time could bring about a change in the state of a system. In view of the great success we have had operating under this assumption there is no reason, this world being as it is, to withdraw or modify that assumption. This prohibition of duration causality is displayed by the fact that the properties of a physical system which are functionally dependent on time do not usually change discontinuously. We regard exceptions to this in one or another of the following two ways. On the one hand we may regard the discontinuous change at time *t* as a random one whose occurrence at that time cannot be explained in causal terms. For instance, this is how one will regard the emission of an alpha ray during a process of disintegration (unless one subscribes to a hidden variable theory of quantum mechanics). On the other hand, we may suppose that there is some more basic continuous process whose occurrence brought about the discontinuous change of state. This is how the hidden variable theorist regards the previously mentioned emission process and how we regard, say, the sudden occurrence of flame in the case of spontaneous combustion. In neither case do we regard discontinuous change as explicable by reference to the causal power of the passage of time.

Like date causality, duration causality *seems* to be a sort of explanatory dead end. For suppose we seek to explain the colour transformations by appeal to the following allegedly law-like regularity. If such and such a state of an object obtains for such and such a period of time, such and such a colour transformation regularly follows. Having posited this regularity as law-like we should naturally wish to account for it in terms of other regularities. In particular we would like to know just why a period of time of that length and not some other length is required. If, contrary to our hypothesis, some process is going on in the period which brings about the change, that may explain just why a period of that particular duration is required. This is how we might, for instance, seek to explain why it takes five minutes of boiling for an egg to become hard. There is a process of heat absorption going on and in normal circumstances it takes five minutes for the required amount of heat to be absorbed. But under our assumption that no process relevant to the transformation is going on during the period of time in question, positing duration causality seems to involve regarding the operation of such causality as a brute regularity not capable of further explanation. Thus, while there is no incoherence in the notion of duration causality, there are reasons for wishing to avoid its postulation. For this reason it might be preferable at any state in an explanation to posit an undetectable process rather than to posit duration causality.

This conclusion is perhaps too strong. For we might envisage having evidence for a number of causal relations of this sort. That is, we might have a sequence of generalizations of the form: 'If conditions C_i obtain for time Δt_i then effect E_i will occur.' It is possible to envisage higher-order generalizations of this form that would entail these lower-order generalizations and hence explain them. The higher-order generalizations might enable us to predict and successfully test a novel lower-level generalization of the form 'Whenever conditions C_a obtain for time Δt_a result E_a will ensue.' Thus, if it were possible to obtain a deductively organized body of such generalizations we would regard the initial lower-level ones as explained and would not have to regard them simply as brute regularities. However, the substantive point remains. As we have no experience with these sorts of regularities it would be methodologically advisable as things stand to try to deal with any apparent occurrences of date or duration causality in terms of theories of the familiar type which would rule out such causality. Only against a background of repeated failures to succeed in this should we be methodologically justified in trying to develop theories involving causal time.

In considering what I called duration causality we supposed that some states obtaining for some period of time could be a sufficient condition for the occurrence of some change. We should also ask for the sake of completeness whether some events occurring or some states obtaining could be a causally sufficient condition for the occurrence of some events at a later time regardless of what happens in the intervening period. Could, in some possible world, the immersion of a billiard ball in water for one minute bring about a transformation in colour one hour later regardless of what happens to the ball in the intervening period (assuming of course that it continues to exist)? This putative form of causality, which might be called *delayed time causality*, is incoherent if it is necessarily true that causes are temporally contiguous with their effects. The issue of delayed time causality will not be pursued here. I would argue that causes need not be temporarily contiguous with their effects and that we can envisage contexts in which it would be sensible to posit duration causality. Of course we would be at least as reluctant to invoke delayed time causality as we would be to invoke duration causality.

Assuming that positing temporal vacua in the fantasy worlds requires us to have some causal explanation of the ending of the vacua and assuming similarly that positing the discrete quantized model of microscopic change considered above requires us to have some causal explanation for the ending of the mini-freezes, these contexts in which it might be reasonable to posit temporal vacua are contexts in which we would be constrained to abandon our belief in the non-occurrence of duration or delayed time causality. While there are clearly good reasons, as we have seen, for being reluctant to make this move, there does not seem to be an incoherence involved in so doing. While we normally think of time as acausal and as linked to change, the link with change is not so firm as to preclude the sensible theoretical positing of time without change. However, this can only be done at the cost of vesting time with some sort of causal power, which for the reasons given, we should be most reluctant to do. Of course, the reductionist, committed to holding *AP* to be a necessary truth, will regard causal time as incoherent. But undesirable as causal time may be, it is not incoherent Reductionism thus fails as an analysis of our concept of time.

The reductionist was characterized as offering a putative analysis of our concept of time. The argument of this chapter established the logical possibility of a temporal vacuum. Consequently the reductionist's claim fails. However, one could still adopt *methodological reductionism*. That is, one could adopt *AP* as a *regulative principle*. For any context that might incline us to talk of a temporal vacuum is a context in which alternative hypotheses could be held in the face of all possible observations. Given *AP* as a regulative principle, one could decide to treat time in a reductionist manner. In this case one would be offering a reformative programme whose goal involves treating time as if reductionism were true.

This reformative programme would have affinities with Quine's thesis that the paradigm case of philosophical analysis is *elimination through explication*. For Quine, in providing a philosophical analysis of an expression we do not in the typical case aim to provide an explication of the meaning of the expression.

> We do not claim to make clear and explicit what the users of the
> unclear expression had unconsciously in mind all along. We do not

expose hidden meanings, as the words 'analysis' and 'explication' would suggest; we supply lacks. We fix on the particular functions of the unclear expression that make it worth troubling about, and then devise a substitute, clear and couched in terms to our liking, that fills those functions.[23]

Among the questions to which Quine seeks to apply explicative elimination is the question: what are numbers? If one seeks as Quine does to answer this question through the use of set theory one is faced with an *embarras de richesse*. For there are any number of different sequences of sets which could be identified with the sequence of numbers 1, 2, 3, ... For instance, one can use the sequence

$[\phi], [\phi, [\phi]], [\phi, [\phi], [\phi, [\phi]]], \cdots$
or the sequence $[\phi], [[\phi]], [[[\phi]]], \cdots$

There is nothing to choose between either sequence of sets.[24] For in either case the numbers so defined will satisfy Peano's axioms for the natural numbers. As Peano's axioms provide the best characterization of the properties of the numbers, insisting the numbers are sets means admitting that we cannot determine which sets the numbers are. Quine's response is that we should cease to concern ourselves with the question: what are the numbers really? We should instead simply adopt whichever sequence of sets is most convenient for the purposes which are at hand and proceed to use those sets as we use the numbers.

That all (i.e., all sequences of sets satisfying Peano's axioms) are adequate as explications of natural numbers means that natural numbers, in any distinctive sense, do not need to be reckoned into our universe in addition . . . will do the work of natural numbers, and each happens to do the work of natural numbers, and each happens to be geared also to further jobs to which others are not.[25]

In a similar vein, a reductionist might recommend setting aside the question: what is time? In its stead he would ask: can we cope with the world without reference to time and temporal items by making do with events and constructions built up out of events? Given as we noted that one can stick by *AP* (albeit at the cost of greater complexity), we should at this stage tentatively answer the question in the affirmative. That is, while the reductionist cannot defend his programme as an analysis of our concept of time, he may be able to defend the view that that concept can be ignored in favour of talk about events. However, if

37

we jettison the pursuit of the truth about our concept of time in favour of the pursuit of the best framework for dealing with the world, where this is to be settled pragmatically by discovering which framework is 'best geared' to do the work at hand, the pragmatic factors indicate that we ought to abandon reductionism in the case of the fantasy worlds we have been considering. For sticking by reductionism in those cases generates the more cumbersome and more complex framework. Thus, the arguments used to show that reductionism fails to provide an analysis of our concept of time, establish that there are contexts in which it would be pragmatically ill-advised to stick by reductionism as a reformative programme for the elimination of the concept of time as opposed to its analysis. In any event it is rash to assume as quickly as Quine does that we cannot come up with an illuminating analysis of a problematic concept. For instance, in his example of the natural numbers, the fact that we have a plurality of set theoretical objects which can equally well serve as the numbers may indicate that we have been proceeding on the wrong tack in our attempts at analysis, and not that we are faced with an unclear, inspecific concept that needs to be purged. We will return to the question as to whether an illuminating analysis can be given in the case of time in chapter X.

For the reductionist, a period of time is necessarily a period during all parts of which something is happening. We have seen that this claim is too strong. While there is some link between time and change, it is not so strong as to preclude the sensible positing of temporal vacua if such a posit provides the best available description and explanation of observable change. We do feel, however, that there is some non-contingent link between time and change, and we should not like merely to conclude that *AP* does not hold in this strong form. We should like to articulate some necessary truth linking time and change that allows for the possibility of temporal vacua. Such a principle which we will now consider can be articulated, and it asserts, roughly, that periods of time are not necessarily tied to events but are necessarily tied to *actual and/ or possible events.*

8 SPATIAL VACUA AND REDUCTIONIST THEORIES OF SPACE

In developing and defending this principle it will be instructive to consider an analogous objection to a reductionist theory of *space*. A reductionist theory of space modelled on the reductionist account of time that we have been considering would involve the claim that all assertions about space are equivalent in meaning to assertions about the spatial relations between items of a certain type. If the items in question are taken to be physical bodies the question arises as to how the reductionists can admit the possibility of a spatial vacuum. We want to be able to say, for example, that although there are no objects between my head and the ceiling there is an extended spatial region between my head and the ceiling. We want to distinguish this situation from, say, there being neither objects between the floor and my feet nor an extended spatial region between the floor and my feet. Intuitively the difference between these cases is the following. While in the former case there are no objects between my head and the ceiling it is *possible* for objects to be between my head and the ceiling without there being any alteration in the spatial relations between my head, the ceiling, and some set of objects serving as the basis of a spatial frame of reference. In the latter case, to suppose that there is some object between my feet and the floor presupposes some alteration in the spatial relation of my feet and/or the floor to the frame of reference. Unless we are courageous enough to deny with Leibniz[26] the possibility of spatial vacua, we must say that a position is *not just where an object is but where an object is or could be*. We can further elaborate this view as follows. We standardly identify a position by giving a description of that position in relation to a collection of bodies serving as a frame of reference F. Let R be such a description. If we say that there is a position bearing R to F if and only if there is some body which bears R to F we are committed to denying the possibility of a spatial vacuum. If on the other hand we say that there is such a position as R if and only if *it is possible* for some object to bear R to F, we can consistently admit the existence of spatial vacua.

If the successful introduction of a term for a spatial position requires the truth of a modally qualified proposition of the above form, what sort of modality is it? It seems that we cannot construe the modality as logical. For suppose in fact that space is finite and unbounded. Hence there is some enumeration, E, of cells of some fixed finite non-zero volume which covers the entire space. We could introduce a description of a position, R, where it is specified that R is not a member of E and R is further from the origin than the length of a path from the origin linking all the cells in the enumeration E. It is clearly logically possible that some object should have this position. For that is just to say that it is logically possible for the space to be larger than it in fact is. Thus, as we want to allow that space could be finite and unbounded we cannot construe the modality in question as a logical modality. Neither, however, can it be construed as a simple physical modality[27] where this notion is understood as follows. It is physically possible that *p* if and only if *both* the conjunction of *p* and the set of all and only those propositions expressing physical laws is consistent *and* the conjunction of this set and the negation of *p* is consistent. To see this, consider the following situation. There might be an unoccupied spatial volume, *V*, such that it is physically impossible for any body to occupy *V*. This situation could arise as the result of the presence of a field which surrounds but does not occupy the volume *V*. The field might be of such a character that it is incompatible with the physical laws that a body should pass through a field of that type. If we consider the grounds we might have in such a situation for asserting the existence of the unoccupied volume, *V*, we will obtain a clue about the correct construal of the modality in question. To simplify the situation let us suppose that the apparently unoccupied volume lies between two objects *A* and *B*. We might discover that we cannot place any object between *A* and *B* without altering their spatial relations to the given frame of reference. In this case, to justify our claim that *A* and *B* are not in spatial contact, we will have to appeal to geometrical assumptions about the space in question. Given a geometry for the space, we can use some technique (for example, triangulation) to measure indirectly the distance between *A* and *B*. A discovery that this distance is non-zero could be offered as evidence that there is an extended spatial region between *A* and *B*. Perhaps we should say that this discovery would only constitute presumptive evidence. For if we were not able to come up with a satisfactory physical explanation of why we cannot place a body between *A* and *B* we might prefer to assume a different physical geometry relative to which *A* and *B* would be said to be in contact. This consideration suggests that we should construe the modality in question as follows. There is a position bearing *R* to the frame *F* if and only if there is no incompatibility in the joint assumption that the space in question has physical geometry *G* and that some object bears *R* to *F*.

This treatment of space as something like the permanent possibility of spatial relations which allows for the possibility of spatial vacua may seem incompatible with the reductionist's central ontological contention that there is no space in the absence of matter. Recently Hooker[28] has urged this objection:

40

For the possibility of certain spatial relations obtaining presumably continues to hold even in the absence of matter. It still seems to make sense to say, in a universe void of matter, 'If there were to be material objects $O_1 \ldots O_n$ standing in spatial relations r_{ij} $(i, j, = 1, \ldots n, i \neq j)$ to one another and if there were to be another material object O_{n+1} standing in spatial relations $R_1 \ldots R_n$ to $O_1 \ldots O_n$ respectively then O_{n+1} would be at position p'' as it does to say similar things in our present situation. If on the other hand the Relationist admits that these possibilities in some physically interesting sense continue to exist even in the absence of matter (or if he attempts to introduce spatial locations as theoretical entities backing his reconstruction of unoccupied spatial locations), then it seems he must admit that space exists independently of matter, i.e. he must abandon his position.[29]

Hooker's claim is that if we allow, in a reductionist (relationist) account of space, reference not only to actual positions (by actual position I mean a position actually occupied by a body) but also to possible positions (by possible positions I mean positions that could be occupied by bodies), this account is not essentially different from that of the Platonist (absolutist) for whom space exists independently of matter. For, surely, it is claimed, the existence of possible positions cannot depend on the existence of actual objects. However, this is too strong. The reductionist can defend what he would regard as his essential tenet — that there is no sense attached to talk of positions except in so far as those positions can be identified with reference to actually existing objects (and hence no sense attaches to talk of empty space) while still allowing reference to possible positions. For he can maintain that what is required to justify reference to a particular possible position is the truth of some modal proposition in which reference is made *outside the scope of the modal operator* to actually existing physical objects. That is, he can employ something like the following scheme:

> There is a position such that R if and only if given some actual set of objects constituting a frame of reference *it is possible* (in the sense explicated above) for some object O to bear R to that frame.

This analysis allows reference to unoccupied positions in a manner that is parasitic on reference to occupied positions. For possible positions must be identified in relation to some actual objects. Thus the reductionist can defend the essence of his ontological position, which is that

41

there is no space in the absence of body, while allowing reference to actually unoccupied but possibly occupiable positions. We will refer to a reductionist account of space that involves use of modalities along the general lines indicated as *modal reductionist theories.*

9 A COMPARISON OF TEMPORAL AND SPATIAL VACUA

Before considering in the next section the application to the temporal case of a line of reasoning analogous to the above, it will be instructive to consider our intuitive predisposition to admit the possibility of a spatial vacuum and our intuitive predisposition to deny the possibility of a temporal vacuum. While the possibility of temporal vacua means that there is no asymmetry between space and time with regard to the possibility of vacua, there is certainly an asymmetry between our willingness to embrace vacua of space and of time which needs exploration. I have suggested that any context in which we would be warranted in positing the existence of a temporal vacuum is a context in which we would have to abandon certain deeply entrenched beliefs concerning the acausality of time. While we do have beliefs about the acausality of space, it seems that we can retain consistently these beliefs along with the admission of a spatial vacuum. We do not think that the spatial location itself of some occurrence can be a causal factor in bringing that occurrence about. The admission of a spatial vacuum in no way calls this belief into question. We also hold that mere uniform motion of bodies (the analogue perhaps of the passage of time) does not bring about any effect over and above change of position. While we might well want to abandon this belief in some contexts, it is not called into question by the existence of spatial vacua. If causal influences seemed to propagate across spatial vacua, we might be led to posit action at a distance. This seems to be the only counter-intuitive consequence of admitting a spatial vacuum. While there is a long history of resistance to action at a distance, the prohibition against postulating action at a distance is not nearly so deep-seated as the prohibition against positing causal time. And in any event not all contexts calling for the postulation of spatial vacua would be contexts calling for the postulation of action at a distance. Thus there are reasonable grounds for being more reluctant to admit temporal vacua than to admit spatial vacua. For the positing of temporal vacua, unlike the positing of spatial vacua, requires revising deep-seated beliefs about the sorts of causality operative in the universe.

42

To elucidate further our predisposition to allow spatial but not temporal vacua one should reflect on the multi-dimensionality of space. We noted above the importance of the possibility of indirectly measuring the distance between bodies in justifying the postulations of a spatial vacuum. In the case of the temporal vacua we did not suppose that we could *measure* its duration either directly or indirectly. We were led to suppose that it had a certain measure on the grounds that supposing it to have that measure would give us simpler hypotheses fitting the observed regularities. Thus, in the case of time or a one-dimensional space, the grounds for asserting the existence of a vacuum will be more tenuous and indirect than in the case of a two- or three-dimensional physical space. For this reason it is less problematic to deny the existence of vacua in time than in our actual physical space.

In considering our predisposition to countenance spatial but not temporal vacua we should note the following. As spatial locations exist through time we can bring it about that some spatial regions become for some time unoccupied spatial regions. We can remove an object from some location, thereby leaving an unoccupied location to which the object can be returned later. We cannot remove an event, leaving a temporal 'hole' to which the event can be returned later. For periods of time do not exist through time as regions of space exist through time. It may seem unsatisfactory to attempt to account for the predisposition in question by reference to the absence of an analogue to the familiar process of creating a spatial vacuum (or, at least, an approximate vacuum). For it may be objected that we have not considered the precise analogue of this process. If we characterize the process in question as that of bringing it about that a location occupied at time t is unoccupied at time t' the precise analogue of this is to be characterized as bringing it about that at one location a period of time is occupied (i.e., something is there occurring) and at another location that period of time is unoccupied (i.e., nothing is there occurring). However, while this is a possible process, it would not constitute the production of a temporal vacuum. For this requires that *no* spatial location is occupied by an event during the period of time. However, in the case of a spatial vacuum, all that is required is that *some* spatial location be unoccupied during a period of time. We can produce approximate spatial vacua because that requires only tampering with a part of space during a period of time whereas producing a temporal vacuum requires tampering with all of the space during a period of time.

43

10 TEMPORAL VACUA AND MODALITIES

Analogous to the spatial case in which a vacuum can be thought of as consisting of an unoccupied region which might be occupied, we might explore the possibility of thinking of a temporal vacuum as a period of time in which nothing happened but in which something might have happened. To elucidate this further, consider the world discussed in section 1 of this chapter and suppose for the moment we are assuming that in the year 1984 there was no temporal vacuum. Let E be some event bridging midnight of New Year's Eve in the year 1984. Let E_1 and E_2 be events which are cuts of that event in the following sense. E_2 is that event which is the part of the event E which begins on the instant of midnight. E_1 is that part of the event E up to that instant. Assuming time to be at least dense there is no last instant in the history of event E_1. Thus we can say that, assuming no temporal vacuum, E_1 and E_2 are adjacent events with no events occurring between them. Suppose now we are led to revise our hypothesis and introduce the posit of a temporal vacuum. By analogy with the schema offered in the spatial case we might say:

There is a period of time between the events E_1 and E_2 if and only if relative to these events *it is possible* for some event or events to occur between them.

As in the spatial case we cannot construe the modality in question as a simple logical modality. For consider some event in this actual world. Let E_1 and E_2 be cuts of that event as defined above. It is logically possible that some set of events should have occurred between E_1 and E_2. That is, there is some possible world like this one except that in its history there is fitted between E_1 and E_2 some sequence of events. But we do not want, in virtue of that, to say that there was a period of empty time between them. Taking the modality as a logical modality would lead us into incoherence. For it is logically possible that some type of event happened, once, twice, three times, etc. between E_1 and E_2. If the logical possibility justified introducing reference to actual periods of time we could say that there was a period, one type E_x event in length between E_1 and E_2; and that there was a period, two type E_x events in length between E_1 and E_2, and so on.

It might seem plausible to construe the modality in question as a simple physical modality. For this seems to fit with our picture of both the fantasy world and the discrete-quantized model of the world.

In the context of that model, if T is some mini-freeze we think that it was purely accidental that no change then took place. If the particles had, so to speak, lined up differently there would have been a change in that interval. In the case of the fantasy world the story as told requires us to suppose that the inhabitants of that world can, by suitably manipulating the value of the crucial parameter, cause objects to disappear for a fixed time period; or prevent objects which would otherwise have disappeared from not disappearing. In this case, any posited period of total vanishing is *ex hypothesi* a time in which something might have existed and changed if the parameter had been suitably manipulated. This also appears to be the case in Shoemaker's fantasy world. For let us suppose that the period of total freeze is identified as the nth year by A's clock system. It is not, as he tells it, a physical impossibility that there should be another region, D, subject to some free cycle that would generate change in the nth year.

Asserting the existence of such and such a temporal vacuum commits us to holding that there might (physically speaking) have been change while in fact there was none. However, the physical possibility of the occurrence of events is not a sufficient condition for the existence of a temporal vacuum. We can see this if we consider again a cut, as previously defined, of some event E into sub-events E_1 and E_2 and suppose some limited indeterminism obtains. That is, suppose it is compatible with all physical laws that E_1 should be followed by E_3 and E_3 followed by E_2. Further, suppose it is likewise compatible with all physical laws that E_1 should have been followed by E_2. However, we should not want to say, unequivocally, that, supposing E_1 in fact to have been followed by E_2, there was some period of time in which E_3 could have but did not happen. In one trivial sense we do want to say this, namely, the period of time identified as the period of time after event E_1 in which event E_2 occurred might have contained instead E_3. E_2 would then occur in the period in which the events that were subsequent to E_2 in fact occurred. However, we should not say that there was a period of empty time in which E_3 might have but did not occur. Talking about such a period of time would be purely gratuitous. What is required, as the case of the fantasy world indicated, is that some improved account of events that actually did occur can be gained by making the posit of a temporal vacua. If we merely have the above sort of limited indeterminism none of our descriptions or explanations of the world is enhanced by talking of the empty period in which things did not but could have happened. In the case of the discrete-quantized model of the world the supposition that there are mini-freezes is not justified by appeal to the fact that something could have changed but did not. The postulation of mini-freezes is partially vindicated by reference to the explanatory force of the theory which entails their occurrence. We have to establish not just that something might have happened between E_1 and E_2 but that there was an empty period of time.

I argued that asserting the existence of a spatial vacuum bearing spatial relation R to frame of reference F is tantamount to asserting that no object bears R to F but that some object could bear R to F. It was claimed that the modality could not be treated as a simple physical modality. For there might be positions which it was physically impossible for an object to occupy. The modality in the case of space has to be relativized to a geometry. To assert that some object could bear R to F is to assert that some object bearing R to F is compatible with the physical geometry of the space in question. As there is no analogue in the temporal case to a physical barrier to the occupation of a spatial region, we can construe the modality as a simple physical modality. However, the truth of the appropriate modally qualified proposition is *only a necessary and not a sufficient condition* for the existence of a temporal vacuum. That there was a temporal vacuum entails that it was physically possible for events to have then occurred. However, as the argument with regard to limited indeterminism established, it is not a sufficient condition.

This account fits with the fantasy world argument and the argument based on the discrete-quantized model of the world. It allows us to give a place to Aristotle's insight by preserving a non-contingent connection between time and change through the introduction of the modal element in question. It does not commit us to any view of time as necessarily existing independently of all things in time. For the modalities in question are not logical modalities and, more importantly, the schema of page 44 requires that there be some actual events in relation to which the temporal vacua can be identified.

We began this chapter by remarking that many writers have thought of *AP* as so obviously necessarily true that they have offered little by way of argument for this view. Indeed the only serious challenge that can be made to this view appears to involve arguments of the type Shoemaker and I have advanced. These arguments suggest a revised principle that retains the connection between time and change which we feel must exist and it is compatible with the counter-examples to *AP*. In this event, I would suggest that the weaker principle does not need further positive argument. It should be allowed to stand in the absence of compelling counter-arguments. My objection to the thesis that *AP* is a necessary truth constitutes an objection to the reductionist position as I have articulated that position. It does not constitute an objection to reductionism if reductionism is defined — as some writers have defined it — as the thesis that all assertions about time and the temporal aspects of things can be parsed as assertions about relations between actual and possible events. I will refer to this latter form of reductionism as *modal reductionism*. The difficulties in this revised form of reductionism will be articulated in section 5 of the next chapter.

III

THE TOPOLOGY OF TIME I: THE LINEARITY OF TIME

If the world may be thought of as a certain definite number of centers of force — and every other representation remains indefinite and therefore useless — it follows that, in the great dice game of existence, it must pass through a calculable number of combinations. In infinite time, every possible combination would at some time or another be realized; more: it would be realized an infinite number of times. And since between every combination and its next recurrence all other possible combinations would have to take place, and each of these combinations conditions the entire sequence of combinations in the same series, a circular movement of absolutely identical series is thus demonstrated: the world as a circular movement that has already repeated itself infinitely often and plays its game *in infinitum*.
Nietzsche, 1968, p. 549

1 THE STANDARD TOPOLOGY

In this chapter we begin our investigation of the topology of time. Initially at least we will think of time as an organized structure of temporal items and we will be concerned to investigate the *topological properties* of this structure. The notion of a topological property is borrowed from mathematics where it is defined as any property possessed by a structure which is preserved under all one–one continuous transformations of that structure.[1] We can bring out the intuitive content of this notion by considering some perfectly elastic object in three-

48

dimensional space that has the topological property of being a torus (i.e., shaped like an anchor ring). No matter how you twist, bend or otherwise distort this object, so long as the surface is not broken or torn, it will remain a torus. For our present purposes we want to contrast topological properties with *metrical properties* which are properties whose explication requires reference to some quantitative measure of distance and are not in general preserved under all one–one continuous transformations. For instance, the volume of a torus is a metrical but not a topological property as it is not preserved if we bend and twist the torus.

I have referred to the notion of a structure. For present purposes we can understand a structure as an ordered set. In the case of time, the set is the set of all temporal items ordered by the various temporal relations defined on that set. An investigation of the topology of time is then an investigation of the topological properties of this structure which I have called the *time system*. We will consider whether time might have the following topological properties among others:

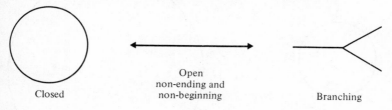

Closed   Open non-ending and non-beginning   Branching

It is easily seen that if we were to represent this structure by perfectly elastic cords the properties in question would be preserved under stretching, contracting and bending, so long as the cords are not torn.

To display fully the contrast between topological and meyrical properties in the case of time, reference must be made to the notion of a metrication of time, a detailed exposition of which is given in chapter VII. A metrication of time provides a systematic way of assigning dates to the beginnings and endings of events, and measures to the durations of events. The dates are assigned so as systematically to reflect the measure of the duration between them. For the sake of illustrating the contrast in question let us suppose that time has a first instant, a last instant and is dense (which is to say that between any pair of distinct instants there is a third). Let us consider five successive events E_1, E_2, E_3, E_4 and E_5, which taken together cover all of time. One possible metrication would assign an equal measure of duration to each event as represented in diagram A.

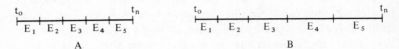

Another metrication would assign greater measures of duration to later events as illustrated in diagram B. We transform A into B by a one–one continuous transformation which has the effect of stretching time out. This stretching does not affect the topological properties of having a beginning, having an end, and being dense. However, if we tried to transform the structure represented in A or B into the closed structures represented in diagrams A$'$ and B$'$ below, we cannot preserve the topology.

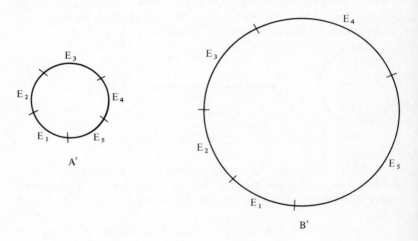

For suppose we curl A up placing t_0 next to t_n. In this case density is lost for there is no instant between t_0 and t_n. If we were to place t_0 next to t_n and then removed one of these instants density would be preserved, but in this case we have violated the prohibition against tearing. Clearly A$'$ and B$'$ on the other hand share the same topological properties for we can smoothly stretch A$'$ to obtain B$'$. In so doing the metrical properties of the events are altered. E_1 and E_5 have the same duration in A$'$ and different durations in B$'$.

An intuitive characterization of the contrast between the topological

50

and the metrical aspects of time has been provided. It is standardly held that there is a philosophically interesting difference between matters of topology and matters of metrication. On the one hand, it is said that the topology of time is uniquely determined. According to Platonists it is determined as a matter of logical necessity and according to reductionism it is determined by empirical facts. On the other hand, it is said, as we will see in chapter VII, that there are no facts, neither of a conceptual nor of an empirical kind, determining a uniquely correct metrication of time. This latter thesis is usually expressed in the claim that the metrication of time is to be established by convention. We will see, on the contrary, that there is at least as much reason to say that the facts do not determine the topology of time in all its aspects as there is to say that the facts do not determine a unique correct metrication.

It is fruitful to distinguish the three following questions. Firstly, what topological structure do we as a *matter of fact think of time as possessing*? Secondly, what is the *full content of our thinking of time* as having such and such a set of topological properties? Thirdly, what is our *justification* for thinking that time has a particular topological structure? In this section and the following two I will have something to say about these questions. The first, unlike the second and third, is easily answered. Most men in the street, practising physicists and philosophers, have thought with the Platonists that time is *like* an unbounded line segment. I will call the topology ascribed to time via this picture the *standard topology*. Since this view of time as having the topological properties of being linear, dense (i.e. there is an instant between any pair of distinct instants), non-ending and non-beginning, is held generally, it provides a convenient starting point for our investigations.

One who ascribes to time, say, the property of linearity, is not committed to thinking of time as an entity or object, either on a par with items such as events which we think of as being in time, or on a par with items such as temporal items which we think of as constituting time; nor is he committed to thinking of time as some *sui generis* 'monstrous object' (Prior's apt phrase) to which we ascribe the property of linearity. There are at least two ways in which we can represent the topology implicit in the standard view which do not carry the suggestion of an objectionable reification of time. One way of achieving this is to use the resources of a first-order quantification language with identity. In this case we take as the domain of quantification the set of all instants or moments of time. We introduce a two-place relational predicate, T, to be interpreted as the relation of being *temporally before*. Any time system in which the relation of being before defined on the set of instants obeys the following axioms will have the standard topology.

T1. $(x) - Txx$
T2. $(x)(y)(Txy \rightarrow - Tyx)$
T3. $(x)(y)(z)((Txy \,\&\, Tyx) \rightarrow Txz)$
T4. $(x)(y)(y \neq y \rightarrow Txy \lor Tyx)$
T5. $(x)(y)(\exists z)(Txy \,\&\, x \neq y \rightarrow Txz \,\&\, Tzy \,\&\, z \neq x \,\&\, z \neq y)$
T6. $(x)(\exists y)(Txy)$
T7. $(x)(\exists y)(Tyx)$

T1, which ascribes to the relation of being temporally before the property of irreflexivity,[2] asserts that no instant comes before itself. T2 (asymmetry) asserts that if *a* comes before *b, b* does not come before *a*. T3 (transitivity) asserts that if *a* comes before *b* and *b* before *c, a* comes before *c*. In conjunction with T3, T2 rules out the possibility that time has a closed structure as represented in Figure A'. For if *a* comes before *b* in such a structure, tracing around the circle gives us by transitivity that *b* comes before *a*, contrary to T2. T4 (connectedness) guarantees that if *a* and *b* are distinct instants, either *a* comes before *b* or *b* comes before *a*. T4 with T1, T2 and T3 ensures that the time system is linear by ruling out, for example, the non-linear structures below, where neither *a* is before *b* nor *b* before *a*. T5 (density) asserts that between any two distinct instants there is a third distinct instant. T6 (non-ending) and T7 (non-beginning) assert, respectively, that there is an instant after any given instant and an instant before any given instant.

The above mode of characterization involves quantification over temporal items and on some views of ontology and ontological commitment we are thereby committed to the real existence of these abstract items. It has thus seemed preferable to some philosophers to characterize what I have called the standard view of time's topology with the aid of the resources of tense logic. In this context the most useful starting point for developing tense logics is Lemmon's system K_t which does not embody any assumptions about the topological structure of time. If one holds that the topology of time is an empirical matter, one will hold that only a formal system such as K_t deserves to be thought of as representing a tense *logic*. One will then regard the so-called tense logics developed by adding to K_t postulates embodying assumptions about the topology of time as systems which go beyond the proper province of logic. With this view I agree. However, for use in exposition I will conform to the standard practice of referring to such systems as tense logics.

Lemmon's system k_t of tense logic is obtained by adding to classical propositional logic the rules G and H and the axioms k1, k2, k3 and k4. Adding to k_t the remaining axioms given below generates a tense logic which attributes to time the standard topology.

Rule G: $\dfrac{\vdash \alpha}{\vdash - F - \alpha}$

Rule H: $\dfrac{\vdash \alpha}{\vdash - P - \alpha}$

K1. $- F - (p \to q) \to (Fp \to Fq)$
K2. $- P - (p \to q) \to (Pp \to Pq)$

K3. $P - F - p \to p$
K4. $F - P - p \to p$

N1. $FFp \to Fp$
N2. $PPp \to Pp$

N3. $Fp \to FFp$
N4. $Pp \to ppp$

N5. $- Fp \to F - p$
N6. $- Pp \to P - p$

N7. $PFp \to (p \lor Fp \lor Pp)$
N8. $FPp \to (p \lor Pp \lor Fp)$

In the above propositional logic the atomic variables $p, q, r, \ldots$ are to be thought of as present-tense propositions[3] such as 'It is now raining' whose truth-value is to be assessed at each instant of time. 'F' ('P') is a one-place sentence-forming operator to be read as 'It will be (was) the case that —.' Any proposition of the form '$F\alpha\,(P\alpha)$' is true at an instant t if and only if there is an instant after (before) t at which α is true. N3 guarantees that time is dense to the future. For suppose that time is not dense to the future. In that case there are instants t_1 and t_2 with no instant between. Let 'p' be true at t_2 and false at all times after t_2. Then 'Fp' is true at t_1 but 'FFp' is not true at t_1 for there is no later time at which 'Fp' is true. Thus, if time is not dense '$Fp \to FFp$' fails. Similarly N4 asserts that time is dense to the past. That N5 asserts that time has no end can be seen by supposing that there is a last instant of time t_1. At t_1 all future-tense propositions are false. For if 'Fp' were true at t_1 there would be a later time t_2 at which 'p' would be true contrary to the assumption that t_1 is the last instant of time. N5 rules out the possibility that all future-tense propositions should be false. For if at t_1 a future-tense proposition is false, i.e., '$-Fp$' is true, by N5 some future-tense proposition, i.e., '$F-p$', is true contrary to the assumption of an end to time. Similarly N6 rules out the possibility that time should have had a beginning. That the set of postulages asserts density and linearity can be established by analogous argumentation.

I shall argue later (ch. VI) that from the point of view of ontological commitment there is nothing to choose between these two modes of clashing claims about the topological structure of time. Consequently, for the purposes of this chapter, I will make use of both methods of explication.

We have now made one step in the direction of explicating the content of the generally accepted claim that time possesses the standard topology. For that claim can be construed as the claim that the given first-order theory is true under the intended interpretation. Or, equally, it can be taken to be equivalent to the claim that the postulates of the tense logic are true at all times, for all interpretations of the atomic variables. This level of abstraction of the content of the claim in question prompts us to think of the possibility that time might have some other structure. It is enlightening in this regard to think of the analogous situation in geometry. For the possession of a suitable axiomatization of Euclidean geometry prompted the investigation of the consequences of modifying those axioms. In this way alternative geometries were produced and were shown to be consistent if and only if Euclidean geometry was consistent. Having available a family of alternative geometries it was no longer natural to assume with Kant that the geometry of the actual world had to be Euclidean. Which geometry we should suppose to be true of this world is to be decided on the basis of a physical investigation of the world. In a similar vein we can tamper with the given first-order theory or tense logic to provide a family of consistent rival theories concerning the structure of time. That is, we can produce a family of consistent theories whose models have differing topological properties.[4] If we accept this analogy with geometry we will regard the question of the topological structure of time as an empirical one which is to be decided by reference to investigations of the physical world. This analogy is accepted by many writers, including Putnam who claims:

> I do not believe that there are any longer any *philosophical* problems about Time; there is only the physical problem of determining the exact physical geometry of the four-dimensional continuum that we inhabit.[5]

Others have followed Kant in arguing that time possesses the standard topology as a matter of necessity. That is, it is claimed that no interpretation of a first-order theory in which the model has a topology other than the standard topology can be thought of as an interpretation in which the relational predicates of the theory are construed as representing temporal relations. Swinburne, for instance, writes:

> Time, being of logical necessity unique, one-dimensional and infinite, has of logical necessity a unique topology. Instants have to each other the neighbourhood relations of points on a line of infinite length.[6]

In so arguing, they embrace the view that time is importantly different from space in this regard for they do not argue that space possesses its topological structure as a matter of necessity. In the course of this and the following four chapters I will argue to the contrary that time could have any one of a number of different topologies. It will emerge, however, that there is an important and interesting sense in which the standard topology of time is privileged (see ch. X, sect. 8).

Explications of the content of hypotheses concerning the topology of time given in terms of first-order theories or tense-logical postulate sets give the *truth-conditions* of these hypotheses. For instance, we explicate the content of the claim that time has no beginning by saying that that claim is true if and only if for every instant of time there is an earlier instant of time, or by saying that that claim is true if and only if for any proposition p, either p is true or it was true in the past that it will be true that p. Such explications will be called *minimal explications.*

Some, including Prior, regard the articulation of a minimal explication as all that needs to be done in giving a full answer to the question of the meaning of the hypotheses in question. Others would be inclined to argue that such an account is not satisfactory unless it is backed up with an indication of what would constitute evidence warranting us in asserting that the truth-conditions were fulfilled and what would constitute evidence warranting us in asserting that the conditions were not satisfied. On this view we have not given the factual meaning of a hypothesis unless we have shown how to recognize whether the truth-conditions are fulfilled. And if the sentence apparently expressing a hypothesis lacks recognizable truth-conditions, it does not in fact express a genuine hypothesis about how things might be. An explication which provides recognizable truth-conditions will be called a *maximal explication* and will be said to provide an account of the assertibility conditions of the claim in question. For the moment we can leave aside the question as to whether any adequate account of the meaning of a hypothesis that purports to be true or false in virtue of how the world is must be maximal. In what follows I will seek to provide maximal explications. Even those who hold that giving the meaning of a hypothesis does not necessarily involve giving a maximal explication would acknowledge that it is of philosophical interest to show that a given hypothesis has specifiable assertibility conditions. Their interest in this will derive from the fact that if a maximal explication can be given for a hypothesis, the hypothesis is shown to be non-vacuous in the sense of being potentially usable. That is, there will be circumstances in which one would be warranted in asserting the hypothesis, and circumstances in which one would be warranted in denying the hypothesis. As an illustration of my distinction between minimal and maximal explications consider the following account given by Prior of the meaning of the hypothesis that time has an end.

> what is meant by time's having an end is precisely that for any *p*, either already it will never be the case that *p*, or it will be the case that it will never be the case that *p* or, to put it another way, that it either is the case, or will be the case, that nothing — not even that such-and-such has occurred — will be the case any more.[7]

Clearly we ought to agree with Prior that it is true that time has an end if and only if it is true that there is a time at which all future tense propositions are false. However, what someone wants to know who asks if time might have an end is what, if anything, would ever lead us to say with warrant that time will have an end. Even someone who rejects the claim that an answer to this query is part of an adequate explication of the meaning of the hypothesis will none the less be interested, as noted above, in the answer to the query. In my terminology, giving this answer will be part of the task of providing a maximal explication of the hypothesis.

2 CYCLICAL TIME AND CLOSED TIME

For the balance of this chapter we will be investigating the possibility that time might be closed or cyclical. Such speculations that time might have a non-standard topology are prompted not only by varying the abstract characterization of the sort given of the standard topology, in the way in which the postulates of Euclidean geometry were varied, but also by our picture-making inclinations. For instance, we sometimes picture time as a wire strung with beads where the beads represent the events, processes and so on that occur in time. This may incline us to ask if the wire might not form a circle. But just what are we thinking when we think that time might be like this? We can easily dispense with one way of taking this picture. One sometimes finds the idea of cyclical time equated with the idea of the same time's occurring again and again *ad infinitum*. The present is pictured as eternally circling the circle, coming back again and again to the same time. This is straightforwardly incoherent. For to entertain the idea of the same time occurring again and again is really to entertain the idea of the same time occurring at different times and that is just contradictory. Times are particulars. Each time occurs only once and there are no two ways about that. If we are inclined to think otherwise it is probably because we are incoherently trying to combine notions of linearity and cyclicality. In order to have the idea of repetition we need to think of a linear ordering of the repeated visitations of the present to some given time. But in that case the times in virtue of being present at different times are different times and we have lost the idea of cyclicality. However, we have still to investigate the possibility that each time occurs only once and that the times are related to one another in the way that the points of a circle are related to one another. To mark this distinction I will refer to the incoherent notion discussed above as that of *cyclical time* and the, at least not obviously incoherent, notion just introduced as that of *closed time*.

We cannot adequately characterize this notion of closed time using the two-place predicate, T, interpreted as standing for the relation of being temporally before. It will facilitate the exposition of this fact if we make the fictional assumption that we are dealing with worlds in which there are only a finite number of instants or moments of time, say four. It will be obvious that the argument is not affected if we replace this assumption by the assumption that, say, there are in fact an infinite number of instants or moments of time. We can represent the hypothesis that time in a four-instant world is open by diagram A, where the arrow indicates that the time at the tail of the arrow is before the time at the head of the arrow.

If we suppose time has a closed structure as represented in diagram B, and attempt to describe that structure using the relation of being before, we will find by tracing around the circle that if *a* is before *b*, *b* is before *a* and, indeed, that *a* is before itself. And this, it might be objected, is just incoherent on the grounds that it is part of what we mean by 'before' that no event or time can be before itself. This is not a telling objection. For one can reply that this alleged fact about ordinary language, if substantiated, would only establish that the ordinary notion of being before is not applicable in the case of closed time. This would be a feeble reply unless it can be shown (as will be done in the next section of this chapter) that there are circumstances in which it would be as reasonable to posit closed time as to posit open time. Positing closed time in these circumstances would require linguistic revisions which would no doubt issue in a distinction between a notion of being *locally before* and a notion of being *globally before*. If an event E_1 is locally before an event E_2, E_2 will not be locally before E_1. E_1 might be locally before E_2, E_2 locally before E_3, . . . E_{n-1} locally before E_n; without E_1 being locally before E_n. In the case of global beforeness, any event will be globally before itself and if E_1 is globally before E_2, E_2 will be globally before E_1. If E_1 is globally before E_2 and E_2 is globally before E_3, . . . E_{n-1} is globally before E_n, E_1 will be globally before E_n. That is, one feature alleged to hold of the ordinary notion of beforeness will be held of local beforeness but not of global beforeness, and another feature will hold of global but not local beforeness.

There is a difficulty in using the relation of being before in the case of closed time, for no two-term relation will be adequate for characterizing order in a closed structure. For any such relation T characterizing temporal order must be transitive, and if the time has the structure of a circle or closed curve, the relation will in that case be both symmetrical (i.e., for all instants a and b, if a bears T to b, b bears a to T) and reflexive (i.e., any instant bears T to itself). A relation having these properties is called an *equivalence relation* and in the case we are considering the equivalence relation will hold between any pair of instants. This means that we cannot use such a relation to distinguish between different orderings of instants or events in closed time. For instance we cannot differentiate between the following two arrangements:

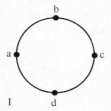 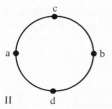

There is a difference in the orderings represented above, but as T will be an equivalence relation it holds between any pair of instants and we cannot describe the difference between I and II using T. A similar line of reasoning will show that no three-term relation is adequate to represent the difference between I and II. However, we can characterize order in closed time if we introduce as our basic temporal relation the four-place relation of pair-separation. We will write '$S(x,y/z,w)$' for 'the pair of instants, x and y, separate the pair of instants z and w'. That is, there is a route from z to w which passes through x but not y or a route between z and w which passes through y but not x.

The properties of the relation of pair-separation can be characterized by the following postulates:

C1. $S(x,y/z,w) \rightarrow S(z,w/x,y)$
C2. $S(x,y/z,w) \rightarrow S(x,y/w,z)$

59

C3. $S(x,y/z,w) \rightarrow$ It is not the case that $S(x,z/y,w)$
C4. $S(x,y/z,w)$ & $S(x,z/y,v) \rightarrow S(x,z/w,v)$
C5. If x, y, z, w are non-simultaneous instants, x is temporally pair-separated from one of the other instants by the remaining two.

We can express the hypothesis that time is closed by adding the axiom

C6. If x, y, z are non-simultaneous instants, and the pair (y,y) separates the pair (z,x), then any pair of these instants is separated by the pair consisting of one of the other instants, taken twice.

The import of this condition can be seen if we consider diagrams A and B. The axiom fails in the case of the structure represented by A because the pair (c,d) is not separated by the pair (a,a). For there is no route from c to d which passes through a. The axiom is satisfied in the structure represented by B as can be seen by an examination of all possible cases. For instance, to take one case, the pair (c,d) is separated by the pair (a,a). For there is a route from c to d which passes through a.

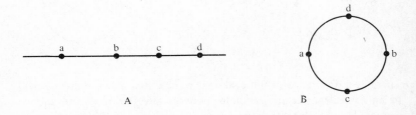

Not only does the relation of temporal pair-separation allow us to distinguish between closed and open time, it is adequate to differentiate between the orders given in diagrams I and II. For in I, the pair (a,c) separates the pair (b,d) and the pair (a,b) does not separate the pair (c,d). In II, the pair (a,b) separates the pair (c,d) and the pair (a,c) does not separate the pair (b,d).

We have noted that no adequate characterization of order in closed time can be given using a two-place relation. A similar problem arises with the tense-logical characterization of closed time given by Prior. Prior[8] adds to Lemmon's basic tense logic k_t the following additional axioms:

$Fp \rightarrow Pp$
$FFp \rightarrow Fp$
$-Fp \rightarrow F-p$

As one would expect, Fp is the provable equivalent to Pp in this system. A consideration of the diagrams below reveals that within this tense logic one cannot distinguish between different orders in which propositions are true in closed time:

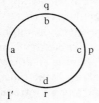

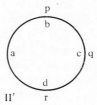

In these diagrams a, b, c and d represent instants of time and p, q and r represent tensed propositions. At a, in both cases Fp, Fq, Fr, Pq, Pp and Pr are true. No complex tensed sentence constructed with these operators can be found which is true at a in I' but false in a in II'.

It is of interest to consider whether a tense logic can be developed which is adequate for the characterization of order in closed time. One reason for this is that some, notably Prior, argue that the primary notions to be used in coming to understand time are those of being the case, of having been the case, of going to be the case and so on. Having argued that the hypothesis of unconnected time-streams cannot be expressed in a tense logic Prior goes on to claim:

> If, as I would contend, it is only by tensed statements that we can give the cash-value of assertions which purport to be about 'time', the question as to whether there are or could be unconnected time-series is a senseless one. We think we can give it a sense because it is as easy to draw unconnected lines and networks as it is to draw connected ones; but these diagrams cannot represent *time*, as they cannot be translated into the basic non-figurative temporal language.[9]

One who took this view might feel that the notion of closed time is problematic on the grounds that the tense logic offered by Prior for closed time is not adequate to represent order in closed time. I do not myself take this view and will argue in the next chapter that the

61

hypothesis of unconnected time-series is not senseless notwithstanding the fact that it cannot be expressed within a tense logic. However, my present concern is not to challenge the general view but to show that a tense logic appropriate to closed time can be developed. The results to be presented are not used subsequently in this work and the reader uninterested in the formalities of tense logic can skip the remainder of this section.

In order to develop an adequate tense logic for closed time we must employ much more complex tense operators than the simple 'It was the case that . . .', 'It is the case that . . .', and 'It will be the case that . . .'. To this end we introduce a three-place future-oriented operator ψ where 'ψpqr' is to be understood as asserting that q and r are sometimes true (at, say, t_q and t_r respectively) and that p will be true at a time t_p which is such that the pair of times (t_0, t_p) (where t_0 is the time of utterance) pair-separate the pair of times (t_q, t_r). Similarly we introduce a three-place past-oriented operator where 'ϕpqr' is to be read as asserting that q and r are sometimes true (at, say, t_q and t_r) and that p was true at a time t_p which is such that the pair of times, t_0 (the present) and t_p, pair-separate the times t_q, t_r. If we consider analogues of the diagrams I and II given above we can see that these operators will allow us to represent order in closed time:

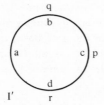

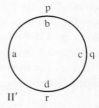

Assuming in I' that p, q and r are only true at times c, b and d respectively 'ψpqr' is true at a and 'ψqpr' is false. In II', assuming that p, q and r are only true at times b, c, d respectively, 'ψpqr' is false at a and 'ψqpr' is true at a.

The intuitively characterized semantics given above for the operators ψ and ϕ can be supplemented by a model-theoretical characterization. In motivating the development of such a semantics, it will be useful to first give a model-theoretical semantics for a simple formal tensed propositional language containing in addition to the usual truth-functional connectives only the simple future-tense operator, F, and the simple past-tense operator, *P*. The atomic formulae *p, q, r*, . . . of such a language are thought of as expressing simple present-tensed propositions. An interpretation of the language will associate a truth-value with *p* for each time. More formally, in specifying an interpretation we first specify a *time structure* which for this simple tense logic will consist of an ordered couple the first member of which, *T*, is a non-empty set thought of as the set of instants of time, and the second member of which, *R*, is a binary relation defined on that set. Relative to a structure, an interpretation for a set of well-formed formulae of the language is a function that assigns to each atomic propositional variable occurring in any formulae in the set a truth-value, either 1 (for truth) or 0 (for falsehood) for each member of *T*. Truth-values are assigned to complex well-formed formulae according to the following recursive rules:

F Rule: '*FA*' is assigned 1 at *t* if and only if there is a *t'* such that *t* bears *R* to *t'* and '*A*' is assigned 1 at *t'*.

P Rule: '*PA*' is assigned 1 at *t* if and only if there is a *t'* such that *t'* bears *R* to *t* and '*A*' is assigned 1 at *t'*.

neg. Rule: '−*A*' is assigned 1 at *t* if and only if '*A*' is assigned 0 at *t*.

vel. Rule: '*A* ∨ *B*' is assigned 1 at *t* if and only if either '*A*' is assigned 1 at *t* or '*B*' is assigned 1 at *t*.

Analogous rules can be specified for the remaining truth-functional connectives. I will call a sentence *tense-logically true* if and only if it is assigned 1 at all *t* for all possible interpretations with respect to all possible structures. A sentence will be said to be *characteristic* of a particular kind of structure if and only if (1) it is not a tense-logical truth and (2) if it is assigned 1 at all *t* under all possible interpretations with respect to any structure of the kind in question.

In the case of a tense language containing in addition to the standard truth-functional connectives the operators ϕ and ψ, I will define a time structure to be an ordered triple $\langle T, P, R \rangle$ where *T* is thought of as the non-empty set of instants. *P* is a four-place relation defined on *T* and satisfying the postulates C1, C2, C3, C4, C5 given on pp. 59–60 for the relation of pair-separation. *R* is a binary relation defined on *T*. Relative to a given structure an interpretation of a set of well-formed formulae or wffs is a function which assigns to each atomic variable occurring in any formulae in the set either 0 or 1 at each *t* in *T*. In giving a recursive definition of truth for complex wffs of the language, the clauses for formulae whose main connective is a truth-functional connective will be the same as those given above for the simple tense language. The clauses for P and F remain as given above. The clauses for ψ and ϕ are as follows:

> 'ψABC' is assigned 1 at t if and only if

(1) there is a time b at which 'B' is assigned 1
(2) there is a time c at which 'C' is assigned 1
(3) there is a time a at which 'A' is assigned 1
(4) t bears R to a
(5) the pair of times (b,c) pair-separate the pair of times (t,a)

> 'ϕABC' is assigned 1 at t if and only if

(1) there is a time b at which 'B' is assigned 1
(2) there is a time c at which 'C' is assigned 1
(3) there is a time a at which 'A' is assigned 1
(4) a bears R to t
(5) the pair of times (b,c) pair-separate the pair of times (t,a).

In the case of closed time, as we noted on p. 58, the binary relation R will be an equivalence relation. That is, R will be reflexive, symmetric and transitive. Symmetric and transitive time structures can be characterized, respectively, by the following theses:[10]

$$-F-P \rightarrow p \quad \text{(reflexivity)}$$
$$F-F-p \rightarrow p \quad \text{(symmetry)}$$
$$FFp \rightarrow p \quad \text{(transitivity)}$$

This means that we can obtain a sentence characteristic of closed-time structures by forming the conjunction of these three sentences. However, one caveat is needed. For there is no sentence which is characteristic of *connected* time structures: that is, structures in which for any two distinct times t_1 and t_2, either t_1 bears R to t_2 or t_2 bears R to t_1.[11] Consequently there is no sentence that rules out the possibility of unrelated systems of times. The class of closed-time structures characterized as above will include structures in which there is a plurality of cycles no time in one of which is related to any time in another.

In any closed-time structure 'ϕpqr' will be true at a time t if and only if 'ψpqr' is true at t. For if time is closed, R is symmetrical with the result that t bears R to a whenever a bears R to t and in that case the clause for ψ and the clause for ϕ give the same results. That this sentence '$\phi pqr \rightarrow \psi pqr$' fails to be true at all times in all open structures can be seen by considering the interpretation represented below in which p is only true at time t_p, q is only true at time t_q and r is only true at time t_r. In that case 'ψpqr' is true at t and 'ϕpqr' is false at t.

As has been indicated, the characterization of closed-time structures does not require the use of the operators ϕ and ψ. However, the point of adding ϕ and ψ to our basic tense logic was not to do this. It was to allow us adequately to reflect order in closed time which, as we noted, cannot be done within a tense logic which contains only the simple operators F and P. Developing axioms for the operators ϕ and ψ in the style of Prior is a straightforward but complicated business which will not be done here.[12] Enough has been said to show that if one wishes to have an adequate tense logic for closed time, one can avoid the deficiencies of Prior's approach through the use of more complex tense operators.

3 A MAXIMAL EXPOSITION OF THE HYPOTHESIS OF CLOSED TIME

We have seen that there is no difficulty involved in developing a formal characterization of closed time, using either a first-order theory or a tense logic. The account provided shows which complex temporal relations and tense operators it would be appropriate to employ, in replacement of those relations and tenses we do in fact employ, if we were to adequately characterize the order of events in closed time. The task of providing a full exposition of the hypothesis of closed time remains. In displaying what might incline us to suppose that time was closed let us begin by thinking as we do that time is linear and open. Let us imagine that we have come to have some evidence that a cyclical cosmological model best fits the universe. So we are led to conjecture that the universe is a sort of giant accordion that always has been and always will be oscillating in and out. Let us suppose further that as the evidence comes it tends to support the bolder hypothesis (a hypothesis of greater content in Popper's sense) that the universe during any one period of oscillation is quite similar to the universe during any other period of oscillation. Finally, someone casts caution to the winds and conjectures: the universe is at t_0 in a state of type S_0. At some time t_1 in the future the universe will again be in the same type of state S_0. Further, following time t_1 the universe will run through a sequence of states qualitatively identical to the sequence of states that it runs through between t_0 and t_1. And so on and so on it runs with boring repetition indefinitely into the future. And, similarly, it has run with this same lack of novelty forever in the past. This conjecture, which will be called theory T_1, is the conjecture that time is linear and history is precisely cyclical.[13]

Under the supposition that all the observations that can be made support theory T_1 we can construct a rival theory T_2 which will be equally supported by that data. This is the theory that *time is closed*. That is, time has a structure like that of a circle so that there is just *one* occurrence of each type of state. Each occurrence of a particular type of state, however, lies in both the past and the future. We can represent the contrast between theories T_1 and T_2 in the diagrams given below:

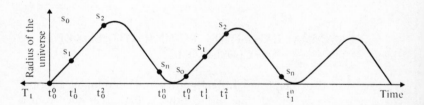

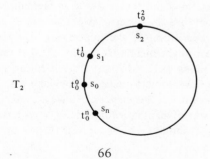

66

The theory T_2 will commend itself to one who — operating under the *initial* assumption that time is linear — reasons as follows. There is nothing that distinguishes the time t_1 in the future when the universe will be in a state S_0 from the time t_0, the present time, at which the universe is in state S_0. For whatever is true of the time t_0 is true of the time t_1. Consequently the best conjecture to make is that the time t_0 is identical to the time t_1: in making this conjecture one is dropping the initial assumption that time is linear and replacing it by the hypothesis that time is closed.

It might appear that there is a third theory which would fit the data equally well. Namely, the hypothesis that time has the structure of an open curve with end-points (i.e., the structure of a closed-closed interval of real numbers) and that each type of state occurs but once with S_0 occurring at one temporal end-point and S_n occurring at the other temporal end-point. However, we are supposing that the laws which are used in predicting and retrodicting future and past states of the universe respectively are such that they invariably posit a state before and a state after any given state. That being so, the only two theories that will fit the data are T_1 and T_2.

T_1 and T_2 are clearly incompatible theories. However, any observation that supports T_1 supports T_2 equally and vice versa. The advocates of T_1 find that the data is exactly what would be expected if T_1 were true. The data in this case consists of support for the conjecture that the universe will in the future be in a state qualitatively identical to the current state. However, this data is exactly what the advocate of T_2 would expect, given the truth of T_2. He too expects that there is in the future a state qualitatively identical to the present state. But, according to T_1 this future time at which the universe is in a state qualitatively identical to the present state is a time distinct from the present time. On the other hand, according to T_2 this future time at which the universe is in a state qualitatively identical to the state it is in at present is the same time as the present time. So, in the imagined context, the choice between theories T_1 and T_2 is empirically undecidable. In other contexts the choice is decidable. For instance, if we have evidence to support a hypothesis of endless novelty (say, that the universe will expand indefinitely) we have grounds for rejecting T_2.

There are two points about these examples to which explicit reference should be made. First, in the story as told there is no possibility of distinguishing between, say, the times t_0^i and t_4^i by reference to differences in the mental states of observers present at these times. For the state of the universe including the mental states of conscious beings are qualitatively identical at these times. We can either suppose that someone present at t_0^i dies before t_1^i and is replaced by a qualitatively identical counterpart (who, depending on which theory we adopt, is or is not numerically identical to the counterpart person) or we can suppose that the person in question has a continued existence between t_0^i and t_1^i, in which case what he claims to remember at t_0^i is exactly what he claims to remember at t_1^i. Second, it is important to note that the two theories in question ascribe non-isomorphic structures to time. That is, there is no one–one order-preserving map (bijection) between a structure having the topology of a closed curve and a structure having the topology of an open curve. If the theories ascribed isomorphic structures to time there would be no genuine incompatibility between them (at least in this regard). For the difference between the theories would amount only to a difference in the mode of characterization of the structure of time and not to a difference in what the structure is supposed to be.

The following heuristic device may bring home with greater force the empirical undecidability of these two theories. Consider two possible worlds, A and B. We stipulate that in world A time is linear and change precisely cyclical so that a theory of the type T_1 is true of this world. We stipulate that in world B time is closed so that a theory of the type T_2 is true of world B. In addition we decree that the entire set of states constituting world B is qualitatively identical to the sequence of states in any one oscillation of world A. Imagine that you are to be placed in one of these worlds without being told which world it is. The question then is − just what possible observation could you make to ascertain whether you are in world A or in world B? My suggestion has been that there is nothing you could do. While these are apparently very different kinds of worlds the question as to which world it is that you are in is empirically undecidable.

4 THE UNDERDETERMINATION OF THEORY BY DATA

I have claimed that in the possible world which we have considered any observation which supports T_1 supports equally T_2 (and vice versa) and that any observation incompatible with T_1 is incompatible with T_2 (and vice versa). As such this constitutes an example of what will be called underdetermination of theory by data. We will see later in this work (ch. VI) that space and time give rise to another example of underdetermination. How one ought to respond to the éases of under-determination is a matter of considerable controversy and, as will be indicated in chapter X, the response one makes largely determines the general theory or picture of time that one adopts. Because of the importance of this notion of underdetermination in the context of theorizing about time, it will be discussed in this section before we return to a further consideration of closed time. First, I offer a characterization of the notion. Second, I will seek to refute the arguments of those who claim that underdetermination cannot arise. The question of what response is appropriate in cases of underdetermination will be deferred until the final chapter.

Quine expresses the thesis as follows:

> Consider all the observation sentences of the language: all the occasional sentences that are suited for use in reporting observable events in the external world. Apply dates and positions to them in all combinations, without regard to whether observers were at the place and time. Some of these placed-time sentences will be true and the others false, by virtue simply of the observable though unobserved past and future events in the world. Now my point about physical theory is that physical theory is underdetermined even by all these truths. Theory can still vary though all possible observations be fixed. Physical theories can be at odds with each other and yet compatible with all possible data even in the broadest sense. In a word, they can be logically incompatible and empirically equivalent. This is a point on which I expect wide agreement, if only because the observational criteria of theoretical terms are commonly so flexible and fragmentary.[14]

That is, we have a case of underdetermination if for some subject matter we have two theories, T_1 and T_2, which are (1) incompatible and (2) both compatible with all actual and possible observations. It is essential to bear in mind the force of the reference to actual and *possible* observation. For the thesis is not the uncontentious claim that a situation could arise in which at some moment of time all observations and experiments made to date have *de facto* left two rival theories in the field. For in the context of the sort of underdetermination Quine has in mind the outcome of *any* possible observation, test or experiment would either support both theories equally or count equally against both theories. If there can be such situations, the question as to which (if either) of the theories is true or correct would be *empirically undecidable*.

According to Quine *all* theories are underdetermined by the data; that is, it is held that for any subject matter there are incompatible theories, all of which fit the data equally well. I will refer to this thesis of Quine's as the *strong UT thesis*. By the *weak UT thesis* I will mean the thesis that there *can be* cases of underdetermination. It is not clear that Quine has any non-question-begging argument to support this strong contention. Indeed, he tends quite candidly to offer this as something on which general agreement can be expected. Not unexpectedly the response to this contention of Quine's has been one of scepticism. This is not to say that such scepticism has been well-grounded by the production of a multitude of forceful arguments. On the contrary, while there has been considerable discussion concerning Quine's claim that the indeterminacy of translation follows from the underdetermination thesis, this thesis itself has received scant attention. No doubt this is so because most writers have been inclined, not unreasonably, to retort that Quine has provided no reason to think the thesis tenable. In this chapter I have partially rectified this deficiency on Quine's behalf by providing a proof by the example of the underdetermination in certain circumstances of the choice between the hypotheses of closed time and open time.

One might wish to object that the thesis as formulated involves the untenable presupposition that there is some viable dichotomy between observational propositions and theoretical propositions. For Quine assumes that a pair of theories are incompatible yet empirically equivalent if and only if they agree on the distribution of truth-values over the set of observational propositions, and disagree on the distribution of truth-values over the set of theoretical propositions. Certainly the assumption that there is a difference in kind between observational propositions and theoretical propositions is dubious. But presumably one who denies that there is such a dichotomy will none the less agree that propositions vary in the degree to which they are observational or theoretical. In this case there will be a spectrum ranging from the more observational to the more theoretical. That being so, we may ask of a pair of incompatible theories where along this spectrum they differ in the ascriptions of truth-values. Whether or not the theories are to be regarded as observationally equivalent and theoretically divergent will depend on where along the spectrum one is willing to say we have become theoretical enough. And as, *ex hypothesi*, we are dealing with a difference of degree and not a difference of kind, this will be a matter for decision. In light of this we can reformulate the underdetermination thesis as follows: No matter where one fixes the point demarcating the observational from the theoretical (so long as one does not put this point at either end of the spectrum) there can be (or, in the case of the strong version of *UT*, must be) theories which agree on the distribution of truth-values to propositions on the observational side of that point and disagree on the ascription of truth-values to some propositions on the theoretical side of that point.

Given that there is no natural bipartite division of the propositions of a scientific language into those that are observational and those that are theoretical, the claim that in a particular case we have underdetermination will have to be relativized to a particular division. Such a claim may be more or less interesting depending on the particular division made. The division implicit in the examples to be given concerning time is such as to render interesting the claim that these constitute cases of underdetermination. In what follows it must be assumed that some division has been made and the claim that two theories are empirically equivalent is to be understood as the claim that the theories agree on the distribution of truth-values over those propositions deemed observational. The question arises as to whether counterfactual observational propositions should be included. For Putnam[15] has objected to Quine's exclusion of counterfactual observation propositions about what would have been the results of experiments that were not but might have been performed (an exclusion which Quine's horror of propositions would lead him to formulate in terms of sentences). This means that two theories which agreed on the truth-values of non-counterfactual observational propositions while disagreeing on the truth-values of counterfactual observational propositions would be regarded by Quine as empirically equivalent and by Putnam as empirically inequivalent. Except for the following comment, this question will not be further discussed, for my example given in this chapter and the example to be given in chapter VI are empirically equivalent on Putnam's more stringent criterion. To see the problem which arises if observational counterfactuals are included, let T_1 and T_2 be theories which are empirically equivalent in Quine's sense and let p be a counterfactual observational proposition which is assigned a different truth-value by these theories. Presumably, if this is to establish empirical inequivalence for Putnam there must be a matter of fact at stake with regard to p. As will be seen later, one response to underdetermination is to conclude that with regard to those propositions responsible for the underdetermination there is no matter of fact at stake. One who makes this response would not think there was a matter of fact at stake with regard to p if the only way to assess the truth-value of p was by reference to these propositions. In view of the fact that the viability of this response is one of the crucial issues about the underdetermination, it would seem reasonable that in assessing empirical equivalence those counterfactual observations whose truth-value can only be assessed by reference to T_1 and T_2 should be excluded.

In considering examples of underdetermination it will be fruitful to consider three strategies that have been deployed in arguing against the possibility of underdetermination. First, it is easy to see that on certain views of the nature of theories underdetermination could not arise. This would, for instance, follow trivially given a reductionist construal of theories which treated all theoretical propositions as translatable into observational propositions. And this result follows almost as trivially if one holds that what a theory really is is more perspicuously represented either by its Craig replacement or by its Ramsey replacement.[16] However, in the absence of convincing reasons for thinking of theories along the lines of Craig or Ramsey these results are of no particular significance. Indeed, one might argue that the possibility of underdetermination provides additional support for the claim that Craig and Ramsey's theories cast no particular light on the nature of theories.

A second strategy would involve arguing that considerations other than those which fit with observational data are relevant in deciding between incompatible theories. An argument of this form is suggested by the following passage from Swinburne's *Space and Time*:

> Compatible with any finite set of phenomena there will always be an infinite number of possible laws, differing in respect of the predictions they make about unobserved phenomena. Between some of these ready experimental tests can be made, but experimental tests between others is less easy and between them we provisionally choose the simplest hypothesis. Evidence that a certain law is simpler than any other is not merely evidence that it is more convenient to hold that suggested law than any other, but evidence that the suggested law is true.[17]

One who thinks with Swinburne that simplicity not only makes more likeable but also more likely to be true might hope to decide between rival theories which are compatible with all actual and possible observations by comparing the theories as to simplicity. Any general appeal to simplicity to rule out underdetermination will be problematic in view of the notorious difficulties involved in producing a reasonable criterion of relative simplicity. And even assuming we had such a criterion there is no reason to assume that *any* pair of incompatible empirically equivalent theories will be such that one is simpler than the other. In the particular case given above and the case to be discussed in chapter VI the appeal to simplicity as a guide to the truth will be irrelevant since on any reasonable criterion of relative simplicity the examples involve theories of equal simplicity.

The third strategy is exemplified in the following remark of Dummett's:

> Quine's argument for the indeterminacy in the strong sense is based on the claim, over which, he says, he expects wide agreement, that there can be empirically equivalent but logically incompatible theories . . . but the claim is absurd, because there could be nothing to prevent our attributing the apparent incompatibility to equivocation.[18]

An adequate discussion of this objection would take us far beyond the confines of this present work. However, we can at the very least show that it is not obvious that the move suggested by Dummett can always be made by constructing examples which it would be unreasonable to treat as cases of equivocation. And perhaps we can do slightly better than this. Suppose we have a pair of theories that appear incompatible but really constitute a case of equivocation. Let T_1 and T_2 be first-order formalizations of these theories (formalizations which stand to the theories as first-order formal arithmetic stands to arithmetic). If we have a case of equivocation, T_1 and T_2 will be mere notational variants of one another in the sense that there are definitional extensions of T_1 and T_2, T'_1 and T'_2, which are such that any theorem of T'_1 is a theorem of T'_2 and vice versa. So if the formalizations of two theories do not have common definitional extensions satisfying these conditions, the theories do not constitute a case of equivocation. Given this as a sufficient condition of non-equivocation, my examples are not equivocal. We will leave this general discussion of underdetermination to consider the implications for the reductionist–Platonist controversy of the possibility of closed time.

5 REDUCTIONISM, PLATONISM AND CLOSED TIME

The reductionist cannot concede the legitimacy of theory T_1. For if he holds that t_0 is distinct from t_1, even though the state of the universe at t_0 is not qualitatively distinguishable from the state of the universe at t_1, he is committed to the view that there can be a difference between times without there being a corresponding difference between the sets of events occurring at the different times. This is a fatal blow to his general thesis that times are certain sorts of sets of events. Neither can he appeal to the numerical distinctness of the event sets in question while allowing their qualitative similarity, for that distinction can only be made if it is conceded that there is a numerical difference in the times.

The reductionist, however, need not simply reiterate his claim that times must be treated as collections of events. For he may, by appeal to some form of the identity of indiscernibles, argue that no matter what view one takes of the existence of temporal items one is committed to identifying t_0 and t_1. Given a suitable interpretation of the principle that if for every property, F, F holds of a if and only if F holds of b, then $a = b$ he will argue that t_0 and t_1 must be identified.[19] An early version of this argument is found in Eudemus of Rhodes:

> Everything will eventually return in the self-same numerical order and I shall converse with you, staff in hand, and you will sit as you are sitting now, and so it will be in everything else, and it is reasonable to assume that time too will be the same.

More recently the argument has been advanced by Grünbaum.[20]

Unless restrictions are placed on the scope of the quantification in the expression of the principle it becomes trivially true. For instance, if one allows within the scope of the quantification the property of being distinct from b, then, given that $a \neq b$, there is trivially a property possessed by a but lacked by b, namely that of being distinct from b. But taken in this form, the principle cannot be appealed to in order to establish the result in question. For we cannot show, without begging the question, that a and b are identical by showing that any property holding of a holds of b and vice versa if in showing that we have to show that the property of being identical with b holds of a. If we take the principle in a stronger form, it could be appealed to by the reductionist in this context, e.g., if we require that the set of properties over which the quantification ranges does not include any property which can only be expressed by reference to the objects a and b. Given this form of the principle the reductionist can argue without begging the question that t_0 and t_1 are the same time.

In making this appeal to the principle of the identity of indiscernibles, the reductionist is denying, in effect, that I have produced a genuine example of the underdetermination of theory. For he maintains that although no rationally grounded choice can be made between T_1 and T_2 by appeal to observable facts, such a choice can be made if we appeal to general principles which can themselves be given an *a priori* justification. That is, while the choice between the theories is *empirically* undecidable, it is not undecidable. For given that the principle of the identity of indiscernibles can be established by *a priori* means to be a necessary truth, the choice is decidable on *a priori* grounds.

The difficulty facing the reductionist is that the claim that the principle of the identity of indiscernibles (in the strong form he requires) is a necessary truth is highly contentious. For it has not been established that accepting the principle in this form is essential to our understanding of identity. And it is just such cases as that of temporally cyclical worlds that have seemed to many to cast doubts on the tenability of the principle. Hence, in the absence of a compelling argument in favour of this form of the identity of indiscernibles, the reductionist case is less than compelling. However, it would be a mistake to insist that the reductionist produce a compelling argument in order to have an interesting thesis about the possibility of closed time. For, as Dummett has argued,[21] nothing prevents one from using this form of the identity of indiscernibles as a regulative principle. If we are not compelled to adopt it, on pain of failing correctly to employ the notion of identity, we are free to adopt it. With this weaker claim, the reductionist cannot compel us to opt for closed time, but he is not precluded from opting for closed time.

The position we have arrived at with regard to reductionism and closed time is, then, the following. A reductionist account of time entails that theory T_1 (linear time and cyclical history) is incoherent. But T_1 is not incoherent. For that could only be established on the assumption that the principle of the identity of indiscernibles is a necessary truth and there is no reason to suppose that principle to be necessarily true. That is, the fact that, in the context of the possible world we have considered, theory T_1 faces no logical inconsistency nor empirical difficulty means that reductionism cannot be regarded as providing a

75

satisfactory analysis of our concept of time. Further, granted, as I have suggested we should, that the principle of the identity of indiscernibles is a regulative principle that one is free either to adopt or to reject, someone can certainly opt for it and adopt theory T_2. But in so doing he cannot charge the supporters of T_1 with any incoherency or vacuity.

The viability of theory T_1 shows similarly that no modal reductionist theory of time can provide a viable analysis of our concept of time. For while it might seem that some form of modal reductionism which sought to construct temporal items on a base that included both actual and possibile events could survive the criticism of chapter II, such a theory is incompatible with the claim that T_1 is a viable theory. No difference arises between corresponding states of the universe in different cycles if possible states or events are introduced.

Since the possible world considered above is a deterministic one, something is possibly the case at time t_i^j if and only if it is possibly the case at time t_{i+1}^j. In such a world we can individuate the corresponding times t_i^j and t_{i+1}^j (which are taken to be distinct in the open-time, cyclic history theory) *neither* by reference to a difference in the actual events which occur at t_i^j and at t_{i+1}^j *nor* by reference to any possible events. In any event it should be noted that any attempt to preserve a form of reductionism in the face of the argument of chapter II by appeal to actual and possible events deprives reductionism of much of its original attractiveness. That attractiveness derived in part from the attempt to reduce apparently problematic and abstract items, temporal items, to less problematic and relatively more concrete items, events. Some philosophers will no doubt feel that the required notion of a possible event is not all that less problematic than the notion of a temporal item.

There are no forceful reasons compelling us to adopt the principle of the identity of indiscernibles. Nor do there seem to be any reasons which ought to lead us to reject it. Consequently, one has a free choice in the context of the possible world described in this chapter — a choice which ultimately is something of a matter of taste. Of course, if the reductionist had vindicated his general view of time, he might urge the adoption of the required form of the principle on the grounds that this is required by the most attractive general theory of time; and, hence, persuade us, in fact, to posit theory T_2. However, as we have seen (chapter II), there are telling objections to the theory as providing an analysis of our concept of time. In any event, to persuade in this way would be to persuade us to adopt reductionism as a methodological

The Topology of Time I

programme and not as an analysis of the concept of time. One who refused to adopt the principle of the identity of indiscernibles and opted for theory T_1 could not be accused by the reductionist of either an inconsistency or a failure to grasp the full content of our concept of time.

If one takes an appropriate formulation of the identity of indiscernibles to be a necessary truth, one will hold that it is an objective matter of fact whether or not time is closed or open. If, on the other hand, one regards it as a regulative principle that one is free to adopt or not to adopt, given an apparently cyclical world history one will be free to treat time as closed or as open. No possible observation will settle the choice between these alternatives. While in this situation it might be natural to operate with a linear temporal framework: in making one choice rather than another we will not run the risk of empirical or conceptual refutation. One cannot, then, say that time either is or is not closed, regarding our choice of one or the other as reflecting our inability to know which it is. For whether or not one regards time as closed, regarding our inability to determine which as reflecting our identity of indiscernibles. But as a regulative principle one cannot say of it that either it holds or does not hold. Hence, one cannot hold, under this general assumption, that as a matter of fact time is either closed or open.

Consequently, it is worth examining a sample of the sort of argument one finds adduced to back up claims that closed time is incoherent. Swinburne has advanced an argument[22] based on the following pair of claims. It is, he alleges, a logically necessary truth that if time t_1 is before time t_2 any state of the universe at t_1 is *unaffectable* by any state at time t_2. It is also a logically necessary truth, according to Swinburne, that any state at t_1 could affect the state at t_2. Thus he concludes that if in positing cyclical time we identify times t_1 and t_2, we are committed to the incoherent proposition that any state at t_1 both could and could not affect any state at t_2.

Swinburne's case rests on the contentious claim that backwards causation is incoherent. We cannot here explore this question. However, in any event Swinburne's conclusion does not follow from his assumptions. For one might grant that in a context in which it really was appropriate to use a directed relation (such as being before) as our basic temporal relation, it is appropriate to use a directed causal relation and maintain that, given these assumptions, if t_1 is before t_2 no state at t_2 can affect any state at t_1 and, holding this, one could deny that it is appropriate in all contexts to use such relations. One might maintain that in positing closed time we are deciding not to employ *these* temporal and causal notions. Given closed time we cannot talk legitimately of events being before or after one another. We must talk instead in terms of the temporal pair-separation of events. Consequently, we will need to employ some weaker non-directed notion of causality. We might then introduce the following notion of causal relatedness. E is causally related to F if and only if, had E not occurred, F could not occur and had F not occurred E would not occur. This seems to be an appropriate notion of causality to use in the context of closed time. One who argues for the possibility of closed time is claiming that there are possible contexts in which we might sensibly posit time to have a closed structure and that such a situation should be described in terms of a causal relation appropriate to that temporal structure. Thus, Swinburne's argument at best gives us the unsurprising result that if one posits closed time and continues to talk in terms of causal notions appropriate only to open time one will generate incoherences.

IV

THE TOPOLOGY OF TIME II: THE UNITY OF TIME

But, what if there are other kinds of things, either different from those about us, or even exactly similar to ours, which have, so to speak, another infinite space, which is distant from this our infinite space by no interval either finite or infinite, but is so foreign to it, situated, so to speak, elsewhere in such a way that it has no communication with this space of ours; and thus will induce no relation of distance. The same remark can be made with regard to a time situated outside the whole of our eternity. But such an idea requires an intellect of the greatest power to try to grasp it; and it cannot be admitted by direct consideration, in any way, or at least with difficulty. Hence, omitting altogether such things, or the spaces and times of such things which are no concern of ours, let us consider the things that have to do with us.
Boscovitch, 1966, p. 199

1 SPACE, TIME AND UNITY

We think of space and time as unities in the following sense. Any particular spatial location is a part of space and all parts of space are spatially related to one another. Any particular time is part of time and all parts of time are temporally related one to another. Kant held that we have no choice but to so think for, he argued, it can be established *a priori* that space and time are unities. Recently it has been argued that Kant was only half right. Both Quinton[1] and Swinburne[2] maintain that while we cannot conceive of time's being non-unified we can describe the circumstances in which we would be warranted in asserting that space was not unified. In this chapter I argue that Kant was right neither about space nor about time. While I think that Quinton and Swinburne have made out their case with regard to *space*, it will prove convenient to offer an argument similar to theirs which will then be modified in the course of developing a case for the possibility of non-unified time.

Besides being of interest in its own right, the question of the possibility of non-unified time (given that non-unified space is a possibility) is of interest for reasons relating to our deep-seated belief that space and time are very different. For it is not infrequently maintained that this belief is not well grounded. A dispassionate investigation of space and time will reveal, it is said, that they really are quite analogous. Consequently, if Quinton and Swinburne were able to make out their case that space but not time can fail to be unified this result would provide one step towards sustaining the general thesis that space and time really are disanalogous. If, on the other hand, my arguments have the force I take them to have, space and time are on a par with regard to the possibility of disunity. An additional but not unconnected reason for being interested in the question of unity relates to the question of the status of assertions about the structure of space and time. Virtually no one today would wish to hold with Kant that we can determine the geometrical structure of space by purely *a priori* argumentation. However, many, including Swinburne, as has been noted, hold that time possesses its topological structure as a matter of necessity and that the character of that structure can be established by *a priori* reasoning. This, I have been and will be arguing, is wrong. The question of the topology of time, like the geometry of space, is an empirical matter requiring *a posteriori* investigation. Thus, to show, contrary to Quinton and Swinburne, that time might fail to be unified will put one further nail in this particular philosophical coffin.

At this juncture some precision must be given to the thesis of the unity of space and time. As a first approximation we might say that space (time) is unified if and only if any pair of spatial locations (temporal locations) bear some spatial (temporal) relation to each other. In order to have an interesting pair of theses to discuss, some restriction has to be placed on the range of this quantification over spatial and temporal relations. For the purposes of this chapter I shall restrict the range of temporal relations to the relation of *temporal connectedness* which is defined as follows. Event e_1 is temporally connected to event

e_2 if and only if either e_1 is earlier than e_2 or e_1 is later than e_2 or e_1 is simultaneous with e_2, where these relations are relativized to an admissible frame of reference. That is, time is unified if and only if relative to any admissible frame of reference any pair of events e_1, e_2 is such that e_1 is earlier than e_2, or e_1 is later than e_2 or e_1 is simultaneous with e_2. In the case of space I restrict the range of spatial relations to the relation of *spatial connectedness* where it is understood that two spatial locations are spatially connected if and only if there is a continuous spatial path joining the two locations. It should be noted that these particular specifications of the general notions of unified space and time would not be appropriate for some types of space and time whose conceivability one might wish to investigate. For instance, the relations of earlier than, later than and simultaneous with are not the appropriate relations if time is closed. As was argued in the last chapter, the appropriate relation to employ in such a case is that of temporal pair separation.[3] One who took time to be closed could express the thesis that time is unified as follows. Any quadruple of events is such that two of those events bear the relation of temporal pair-separation to the other two events. One might wish for some purposes to have more sophisticated characterization of these specific versions of the general thesis utilizing the resources of four-dimensional spacetime geometry. For ease of exposition this will not be done here as nothing in my argument is affected if such characterizations are used.

Consider, for the sake of argument, the putatively possible non-standard topological structures for time represented by three diagrams:

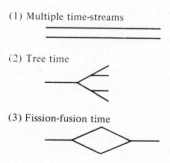

(1) Multiple time-streams

(2) Tree time

(3) Fission-fusion time

The first diagram represents the putative possibility of what I will call *multiple time-streams*. In the case of multiple time-streams there are two or more sets of events, such that any event in one of the sets is temporally connected to any other event in that set and is not temporally connected to any event in the other set. That is, the relation of temporal connectedness generates a partition of the set of all events into equivalence classes (two in the case illustrated). The second diagram represents the putative possibility of time's possessing what will be called a *tree structure*. In this case the events after any node in the tree are later than the event constituting the node and, consequently, later than any event earlier than that event. Any pair of events on different branches or routes through the tree are neither earlier than, nor later than, nor simultaneous with each other. The third diagram represents what we might term *fission-fusion time*. In this case we have two sub-sets of the set of all events such that events in either set have common events in their future and common events in their past. However, no event in either of the sub-sets bears the relation of temporal connectedness to any event in the other set.

On the interpretation I have given of the thesis of the unity of time, all three diagrams are putative representations of ways in which time might fail to be a unity. While I think all three diagrams represent genuine possibilities I will restrict myself to arguing this with regard to the case which is the most outlandish, namely, that of multiple time-streams.[4] The following strategy will be used in establishing this. A conceptually unproblematic description will be given of a possible world and it will be argued that in the context of such a world one would be warranted in advancing the hypothesis that time in this world was non-unified. The point of this strategy is to give a maximal explication of the sense of the hypothesis and so defeat those arguments against multiple time-streams in which it is not argued that the hypothesis involves a contradiction but in which it is argued that nothing could ever constitute evidence in favour of such a hypothesis.[5] I shall not be concerned with the further step in the argument that it is incoherent to suppose that a hypothesis representing a genuine possibility should be unverifiable. I shall be arguing as I did in chapter II that if what is taken as constituting possible verification for a hypothesis is construed in a sufficiently liberal vein to render a verificationist principle of significance even *prima facie* plausible, it is possible to have evidence for the hypothesis of multiple time-streams.

An argument of the above type, which I call an *argument by fantasy*, is of the same general character as that employed in the discussion of *AP* in chapter II. It is designed to show that our concepts as currently constituted admit of a type of application that has been overlooked. As such, the argument is intended to be persuasive rather than knock-down. If it is successful we will learn something about our concepts of space and time by seeing that in certain contexts we would wish to advance hypotheses which at first glance are apt to seem senseless or vacuous. This is not the only style of argument that could be advanced to achieve the same end. One might also proceed by producing a viable physical theory that entails that non-unified time and non-unified space are physical possibilities. While I would maintain that a philosophically adequate understanding of space and time can only be achieved through a consideration of physical theory, there does not seem to be any viable physical theory that is relevant to this particular issue. For this reason I will not here consider physical theories relevant to our understanding of space and time. In any event, we can learn something conceptually interesting by showing that some putative possibility is a genuine possibility by a non-technical route even if a technical route to this end is available. For Swinburne and Quinton have been concerned to argue that our *ordinary* concepts of space and time as presently constituted preclude non-unified time. My aim is to show that non-unified time is a genuine verifiable possibility. In developing my case I will proceed as follows. In the second section of this chapter the possibility of a spatial analogue of multiple time-streams will be established by a fantasy argument. In the following section (section 3) certain parameters in the story of section 2 will be modified to convert that account into an account of a possible world which involves *both* multiple space and multiple time-streams. Section 4 will deal with the considerations which have been advanced for thinking that time must be unified. In general those who have so supposed have made a plausible but untenable assumption about the essentially temporal character of evidence. In the final section we will consider the possibility of developing arguments to establish the possibility of unified space and non-unified time.

2 NON-UNIFIED SPACE

A spatial analogue of two-stream multiple time would be the following. There are two systems of spatial locations, A and B. Each member of A is connected by a spatial path to each other member of A and is not so connected to any member of B, and vice versa. To see what such a world would be like let us imagine we inhabit the following possible world. In this world which is in many ways like the actual world, it is always sunny, warm, and there is an abundance of the good things in life. Let us call it *Pleasantville*. It is a sphere set in an otherwise empty space. It is discovered that eating the root of a certain plant that abounds in Pleasantville causes the eater literally to disappear. Fortunately, after a time, the person reappears. Or, if you prefer, a person very similar both in terms of appearance and in terms of memory impressions suddenly appears. On appearing the person claims to be the person who disappeared and gives us an account of how he suddenly found himself in very unfamiliar surroundings. Indeed, the world as he describes it is cold, cloudy, and the good things in life are scarce. It is, unlike Pleasantville, a cube set in an otherwise empty space. He claims that after living for a time in this world, which he calls *Harshland*, he suddenly found himself back in Pleasantville. In time the consumption of the root in question becomes something of a national pastime. Persons who eat the root together return together and tell stories that are not only internally coherent but mutually supportive. They claim to have found themselves together facing the challenges of Harshland. Indeed there is virtually complete consilience between all the accounts offered of Harshland. And with the exception of the root-eaters themselves, each object in Harshland is quite distinct in appearance from any object in Pleasantville.

Some inhabitants of Pleasantville whom we will call the *radicals* take the line that root-eating provides a character-building opportunity to experience the challenge of life in a harsh evnironment. They organize what they call expeditions to Harshland and encourage the young to participate. In time the radicals even describe themselves as having colonized Harshland. For some of them consume a quantity of the root sufficient to produce what they describe as a visit to Harshland lasting long enough to propagate offspring in Harshland. The offspring of Harshland unions are unable, let us suppose, to visit Pleasantville. Another group of Pleasantville inhabitants, a minority group, whom we will call the *conservatives*, regard the eating of the root as a degenerate form of escapism. They take the line that the drug simply causes people to disappear for a time and to reappear complete with delusory memory impressions. The conservatives consequently only rarely indulge in this frivolity. As one might expect, they are somewhat puzzled about the location of persons who have disappeared. Consequently, they undertake to investigate the possibility that the drug causes the eater to be mysteriously transported to some other region of the universe containing Pleasantville. Happily this universe is a finite and unbounded universe and the conservatives are able to inspect the entire universe and discover that Harshland is not to be found in the universe. So the conservatives are led to suppose that eating the root causes the eater to cease existing for a period of time and to acquire delusory memory impressions.

At this juncture it will be fruitful to articulate some conditions that any acceptable further description of this possible world should satisfy. First, given that the stories told by the inhabitants of such a world display a high degree of internal coherence and inter-personal consilience we ought, all things being equal, to accord their stories an objective status: we should avoid any hypothesis that involves supposing that the root-eaters are merely suffering massive delusions. Instead we should adopt hypotheses which allow us to suppose that their stories are stories of occurrences that really did happen. I will call this constraint on any admissible hypothesis the *objectivity constraint*. Further, given that we have a choice between differing hypotheses, all of which satisfy the objectivity constraint, we should select the hypothesis (should there be one) which minimizes the degree of revision required in the system of beliefs of the inhabitants of the possible world. For instance, if we can meet the objectivity constraint either by supposing that both time and space are non-unified or by supposing that space is non-unified but time is unified we should, all other things being equal, prefer the latter hypothesis. This constraint will be called the *simplicity constraint*.

Adopting the objectivity constraint requires rejecting the hypothesis of the conservatives. The radicals advance a hypothesis that is compatible with the objectivity constraint, namely, the hypothesis of non-unitary space. That is, the radicals claim that there are two complete spaces: the contents of either of these spaces bear no spatial relations to the contents of the other space. While they concede that there may be other hypotheses which satisfy the objectivity constraint, they maintain their hypothesis to be the best bet given the simplicity constraint. Among the alternative hypotheses satisfying the objectivity constraint is one that preserves the unity of space at least in a minimal sense. For it might be held that these two spaces are in fact two regions embedded in some four-dimensional space between which it is physically impossible to travel by conventional means. If no physical theory is forthcoming that *both* explains why it is impossible to travel between the regions *and* specifies the spatial relation between the regions, the hypothesis will not have anything going for it. Moreover, it gives a purely formal sense to the unity of space. Kant, in maintaining the unity of space, was maintaining that space had to be conceived of as a unity in a stronger sense than that involved in this hypothesis of embedding. For Kant also held that space had to have three dimensions. In any event, this alternative hypothesis is just that — it is an *alternative* to the hypothesis of two distinct spaces. With regard to this point it does not matter if I am wrong in maintaining that the hypothesis of two spaces is to be preferred to the hypothesis of two separated regions embedded in a unitary four-dimensional space. For the fact that we have a choice between hypotheses here, and are not compelled in the circumstances to opt for the one that retains some semblance of spatial unity, means that the Kantians are incorrect in claiming that we have to conceive of space as a unity.

In order to make plausible the hypothesis that this story is a story of non-unified space in unified time the story needs further elaboration. Let us suppose that in both Harshland and Pleasantville physically similar systems are used to time events, say, digital calendar clocks. The last reading that a radical sees on a Pleasantville clock on eating the root is the same reading that he sees on first realizing that he is in Harshland. In addition, with regard to processes such as boiling an egg which *seem* to the radicals to take roughly the same length of time in both Pleasantville and Harshland it is found that the number of Pleasantville time units (i.e., as measured by the clocks there) is the same as the number of Harshland units (i.e., as measured by the clocks there). Consequently the radicals have an entirely natural non-arbitrary way of establishing a common temporal metric for both spaces. In a perfectly consistent fashion the events of both spaces can be mapped into a common time by assigning to the events occurring in Harshland the date as measured by Harshland clocks and by assigning to the events occurring in Pleasantville the date as measured by Pleasantville clocks. I will express this feature of the possible world in question by saying that there is an appropriate, non-arbitrary *principle of temporal cross-identification.* In order to amplify what is intended by this, let us suppose for ease of exposition that we are not interested in any finer identification of times than can be achieved using the digital clocks. Let us further stipulate that any period of time in Harshland is identifiable as either the period of time during which a suitable digital clock displayed a particular reading or as the period of time during which the clock displayed some particular sequence of readings. In addition it is stipulated that a similar condition holds with regard to periods of time in Pleasantville. Given that we have identified a period of time in Harshland in this way, to possess a principle of temporal cross-identification is to possess a criterion which allows us to determine in the manner outlined that period of time in Pleasantville (and vice versa). In the story as told, we simply ascertain the period of time in Pleasantville time during which the Pleasantville clocks displayed the same reading or the same sequence of readings that the Harshland clocks displayed during the period of time in Harshland time.

The preceding fantasy argument establishes that non-unified space is a possibility. Those who have failed to see this have simply not been imaginative enough, for with sufficient ingenuity we can give a description of kinds of experiences that would incline us to invoke a hypothesis of non-unitary space. Thus we see that our concept of space as presently constituted does not compel us to think of space as unified. Of course, one might wish to strengthen the case for saying that the fantasy world in question is a world with non-unified space by elaborating the description of the world in various ways. I think it is convincing as it stands. But in any event, enough has been done to put the ball back in the court of one who thinks that space must be conceived of as unified.

Hollis has argued that Quinton's myth cannot be dealt with simply by invoking only non-unified space.[6] According to Hollis, non-unified time must also be invoked. In arguing for this claim Hollis says: 'Quinton assumed travel between the worlds to be instantaneous; but I see no good reason for accepting that it is.'[7] Hollis then proceeds to tamper with certain aspects of Quinton's myth to produce a myth somewhat like the one I advance in section 3 of this chapter. Hollis's modified myth may well constitute an account of non-unified time and non-unified space. It cannot possibly show that Quinton is committed to non-unified time in opting for non-unified space. For possible worlds have just the properties that their describers give them and Quinton's possible world is one in which all 'travel' between the worlds is 'instantaneous'. Quinton need only say that the things that happen in Hollis's myth that incline Hollis to describe it in terms of non-unified time simply do not happen in his (Quinton's) myth.

3 NON-UNIFIED SPACE AND NON-UNIFIED TIME

An account of what it would be like for both space and time to be non-unified can be generated in an economical fashion by modifying our previous story so as to drop those features of the story that allowed us to adopt a principle of temporal cross-identification. The story will become even more outlandish and that is significant. For while Kant, Quinton and Swinburne are incorrect in claiming that we must conceive of time as a unity, it is none the less true that relative to our system of beliefs, non-unified time is a much more outlandish possibility than non-unified space.

To simplify the task of exposition attention will be limited to two inhabitants of Pleasantville, Icabod and Isabel. The reader can at his leisure elaborate the story bringing in other Pleasantville radicals. The salient modifications in the previously given story are the following. There are already inhabitants in Harshland who are unable to 'travel' to Pleasantville. Of course, strictly speaking we should proceed as we did before and tell the story first in a non-question-begging form by relating the impressions that the root-eaters claim to be under on reappearing. However, exposition will be easier if we allow ourselves this licence. Icabod and Isabel are contemporaries in Pleasantville. However, the reports they make on reappearing are of entirely different temporal periods in the history of Harshland. Icabod and Isabel agree in their general description of the large-scale physical features of Harshland. For instance, they both agree that it is a cubical world that is cold, cloudy and unhospitable, etc. Icabod's Harshland is at a stage of development not unlike that of Victorian England. For example, the inhabitants are just discovering the internal combustion engine. Isabel's Harshland is at a stage of development somewhat like twentieth-century California complete with all the marvels of the technological era. Interestingly, she hears stories from the Harshland inhabitants of the mysterious appearances of an Icabod-like character at some point in the past. Further, succeeding visits (relative to Pleasantville time) for Icabod are visits to successive Harshland times within the same general period of Harshland history. Similarly, successive visits for Isabel (relative to Pleasantville time) are visits to successive times (relative to Harshland time) within the same general period of Harshland history. We can represent these mysterious goings-on in the following diagram. The numerals represent the order in which it seems to our protagonists that they have experiences of Harshland and of Pleasantville.

In the revised version of the story we have preserved those ingredients that require the postulation of multiple spaces. However, as will become clear, we cannot treat this as a case of unitary time. For there is no natural non-arbitrary way of cross-identifying the times of events in the two spaces. For if Icabod and Isabel consume the potent root

Harshland	Victorian Harshland				Twentieth-century Harshland			

Harshland ──→

| 2 | 4 | 6 | Icabod | | 2 | 4 | 6 | Isabel |

| | | | | 1 | 3 | 5 | 7 | Icabod |

Pleasantville ──→

| | | | | 1 | 3 | 5 | 7 | Isabel |

simultaneously they will agree on the last clock reading they see in Pleasantville before disappearing. But they report *different* times as the first times they notice on the clocks in Harshland. Thus, we cannot obtain a consistent mapping of the events in both Harshland and Pleasantville into the same universal time by taking it that events experienced by someone just prior to disappearance are just before the events experienced on appearing in Harshland. And a principle of cross-identification of times based on the experiences of some one person would be arbitrary. For there is no reason to prefer an identification based on some one person's reports to that based on some other person's reports. In the absence of any reason to assume that one principle of cross-identification is better than any other principle, it seems reasonable to suppose that there is no common time-system in which the events of both Harshland and Pleasantville can be mapped consistently.[8]

Suppose someone insisted that there must be a common time. This assumption has an interesting consequence. Suppose that the principle establishing the common time maps the time of Isabel's first appearing in Harshland into a time in the neighbourhood of the time of her first disappearance from Pleasantville. Then such a principle will map the first time of Icabod's appearance in Harshland into a time long before the time of his first disappearance from Pleasantville. So, Icabod is a time-traveller. If the principle in question does not map Isabel's times of appearance and disappearance in the manner just outlined she will be a time-traveller. Thus, any attempt to describe this world in terms of a non-unitary space and unitary time has the consequence that must also invoke the hypothesis of time-travel.[9] If, as I would argue, time-travel is conceptually possible, we see that someone who insists on unitary time and non-unitary space is committed to the hypothesis of time-travel and this seems no simpler than the hypothesis of one who opts for non-unitary time and non-unitary space. Given this and the fact that the choice of any particular principle of cross-identification of times is purely arbitrary, there is no reason to favour a hypothesis involving both unitary time and time-travel over a hypothesis involving only non-unitary time. The only reason that could be advanced is the question-begging claim that time must be unitary.

4 THE ESSENTIALLY TEMPORAL CHARACTER OF EVIDENCE

It has been argued that we can envisage contexts in which we would be warranted in holding that time was not a unity. The onus is on me to explain why many philosophers have erroneously believed to the contrary. The reason is that it has been held that evidence is essentially temporal in a sense in which it is not essentially spatial and that that precludes non-unified time being a real possibility, while allowing non-unified space to be a real possibility. The content and the plausibility of this claim about evidence can be brought out as follows. Suppose I have a vivid memory impression of having once gone to an island of such and such a description. In regarding the memory impression as constituting some evidence for the existence of a spatial location containing the island to which I went, I do not commit myself to any belief about the spatial relationship between that location and my present location. However, in so regarding the memory impression, I do commit myself to the belief that the event (my going to the island) is before the present time. Quinton, for example, takes it that there is some such special relationship between time and evidence that precludes non-unified time. After arguing for the possibility of non-unified space he adduces the following consideration with regard to time:

> The moral of these unsuccessful attempts to construct a multi-temporal myth is the same in each case. Any event that is memorable by me can be fitted into the single time-sequence of my experience. An event that is not memorable by me is not an experience of mine ... if an experience is mine it is memorable and if it is memorable it is temporally connected to my present state. The question we are raising — is it conceivable that we should inhabit more than one time — answers itself. For what it asks is: could my experience be of such a kind that the events in it could not be arranged in a single temporal sequence? And it seems unintelligible to speak of a collection of

events as constituting the experience of one person unless its members form a single temporal sequence.[10]

Two radically different theses are suggested by this passage. On the one hand there is what I will call the thesis of the *essentially unified character of personal time*. This is the claim that I am committed to ascribing a single linear order to any *experiences* I claim to remember having had. That is, if e_1 and e_2 are experiences I claim to remember having I am committed to holding that either e_1 and e_2 were experienced together or that e_2 was experienced before e_1 (i.e., that I seem to remember remembering e_2 while experiencing e_1) or that e_1 was experienced before e_2. On the other hand there is what I will call the thesis of the *essentially unified character of public time*. This is the thesis that if I claim to have memory impressions of the occurrences of certain events I am committed to ascribing a single linear order in public time to the events of which I claim to have had experience. That is, if I claim to remember the occurrence of *events* E_1 and E_2, I am committed to holding either that E_1 was simultaneous with E_2 or that E_1 was earlier than E_2 or that E_1 was later than E_2.

The first principle has some plausibility. It is not, as I shall argue below, incompatible with non-unified time. The second principle amounts simply to the denial that a person could have experience of events in different time-streams. Consequently I will not discuss it further. For my argument by fantasy was designed to show that it is indeed possible to be justified in ascribing a non-connected order in public time to events of which one had experiences. And, in any event, Quinton's reference to the 'time-sequence' of my experience suggests that he has something like the first principle in mind.

The thesis of the essentially unified character of personal time amounts to the claim that any subject of experience can attribute a unified order to the experiences he seems to remember of the basis of the character of his apparent memories. While I think one can envisage possible exceptions to this thesis, it is plausible to regard it as some sort of conceptual truth, and I will assume for the sake of argument that it is such a truth. My two-time, two-space myth is quite compatible with this principle. Suppose Icabod and Isabel in their twilight years write autobiographies revealing how it seemed to them. In so doing they will be employing this principle in organizing their recollections. In ordering their remembered experiences on the basis of the character of their memory impressions they will produce an account in which recollections

91

of life in Pleasantville are interspersed with recollections of life in Harshland. However, Icabod and Isabel will not do something that we standardly do, which is to infer that the order we ascribe to our experiences reflects the order of the events in public time of which those experiences are experiences. For as we saw in section 3, if Isabel and Icabod do this they will produce different and incompatible claims about the order of events in public time. If each assumes that time is unified and that the order they ascribe to their experiences reflects the order in a unified public time of which they have experience, there will be a breakdown of a systematic character in the sort of agreement that standardly obtains concerning the ordering of events in public time. If, however, they adopt the hypothesis of multiple time-streams they will be able to pass from the orderings they ascribe to their experiences to an agreed temporal ordering involving two time-streams of the events of which those experiences are experiences.

To reiterate the above point — Isabel and Icabod attribute connected ordering to their experiences on the basis of the character of their memory impressions. Initially, taking time to be unified, they are led to produce incompatible accounts of the ordering in unified time of the events of which they claim to have had experiences. *Ex hypothesi* there is no reason to prefer one of these accounts over the other. Isabel and Icabod can consistently regard both accounts as having objective status if they posit multiple time-streams. Both can then agree on the ordering of events in this non-unified public time. And, in addition, they can thereby explain why their initial judgments (based on the character of their experience) concerning the order of events were incompatible. Thus, even if we agreed to adopt the principle of the essentially unified character of personal time, it does not follow that we are committed to the principle of the essential unified character of public time. Our pre-reflective conviction that time must be a unity no doubt arises from our provincial habit of projecting the unified character of our temporal experience onto the world. In so doing we forget that certain systematic divergences from person to person with regard to the order they ascribe to their experience could constitute evidence for a non-unified ordering of events in public time.

5 UNIFIED SPACE AND NON-UNIFIED TIME?

Could we modify the last version of our basic myth so that it constituted an account of unified space and non-unified timé? That is, could we preserve those features of the myth that precluded the cross-identification of times across the 'worlds' and modify other features of the myth so as to generate a non-arbitrary principle for the cross-identification of spatial locations? With regard to the particular myth in question the answer, it seems to me, is no.

In elaborating this point, let us consider the sort of modifications which would be required in the story of Harshland and Pleasantville to produce an account of a world in which we would be warranted in holding space to be unified and time to be non-unified. If we are to justify the claim that a description of Harshland space is a description of the same space as the space described in the description of Pleasantville we will have to justify the claim that some of the bodies in Harshland are numerically identical to some of the bodies in Pleasantville. If we modify the descriptions of Harshland and Pleasantville to achieve this end, a full description of the history of any modified Harshland will differ from a full description of the history of any modified Pleasantville. For if the descriptions do not differ we have no reason to reject the hypothesis that these are descriptions of the same world in the one unified space and time. But if the full descriptions differ in this way, any body described in a full account of Harshland will be assigned at least some relational properties that will not be assigned to any body described in a full account of Pleasantville. That is, to justify the claim that Pleasantville is Harshland space, we will have to justify the claim that some object given description O_P in a full account of Pleasantville is the same object as some object given description O_H in a full account of Harshland, notwithstanding the fact that the objects O_P and O_H do not share all their properties. This might seem a telling objection to this myth. For we seem to be violating the principle of the indiscernibility of identicals.[11] However, bearing in mind that we are considering the possibility of unified space and non-unified time there is a simple and natural modification in the principle of the indiscernibility of identicals that will preserve consistency. To see this we need only note that no supporter of this principle allows as a counter-example to the principle the fact that an object can have the property F at time t_1 and then lack that property at a later time t_2. For the principle is implicitly relativized at times. Thus, the fact that the object O_H has a property F that the object O_P lacks is not incompatible with the indiscernibility of identicals, for the object in question does not have these incompatible properties at the same time. The time at which the object has F is different (i.e., in a different time-stream) from the time at which the object lacks F. Thus, there is no cogent objection based on an argument from the principle of the indiscernibility of identicals against the possibility of two times and one space. However, the fact remains that there would be no *prima facie* case for the identification of objects across the 'worlds' unless these worlds were very similar in certain respects. But if the worlds are made sufficiently similar we may be able to cross-identify times. For instance, suppose that the object O_P has an atomic structure similar to that of objects in the actual world. Then, unless the object O_H shares that internal structure with the object O_P, there is no case for equating the object O_H with the object O_P. But if there is a similarity between the atomic constituents of the object O_H and the object O_P, this will provide for a common clocking system. In that case we would have a non-arbitrary principle of temporal cross-identification. But this is inconsistent with the principle of the indiscernibility of identicals. For if there is just one time system then at one and the same time the object O_H and the object O_P will be ascribed incompatible properties. Thus, it seems that whatever would warrant us in supposing space is unified would warrant us in supposing time is unified. Consequently, our only hope of satisfying the objectivity constraint lies in adopting either a hypothesis of non-unified space and unified time or a hypothesis of non-unified space and non-unified time.

A similar problem arises with an attempt by Swinburne, who argued that the similarity in gross macroscopic features was adequate to establish the numerical identity of objects described in the differing histories.[12] Suppose both the Harshlanders and the Pleasantvillers describe themselves as living on a similar sphere rotating around a similar sun in an otherwise empty universe. If we make the descriptions of these 'worlds' sufficiently similar to justify saying that it is the same sphere and the same sun being described in both accounts, we will have made the worlds similar enough for there to be a non-arbitrary principle of temporal cross-identification. The 'clock' could be based on the rotation of the sphere around the sun. In this case the objection developed above applies with equal force.

The preceding considerations do not show that we cannot tell a coherent story about what it would be like to inhabit a world with unitary space and non-unitary time. The point that has been made is simply that if we are to be warranted in describing a fantasy world as having a unitary space some objects described in the differing histories must be described in sufficiently similar ways as to justify the claim that these objects are numerically identical, and if this is done it *may* provide grounds for adopting a non-arbitrary principle of temporal cross-identification. In this case the story would fail to provide an account of a unitary space world with non-unitary time. To describe a world with unified space and non-unified time one might try to characterize the spatial contents of worlds which were sufficiently similar to justify the claim that some objects in the 'Harshland' description were numerically identical to some objects in the 'Pleasantville' description. The objects in these 'worlds' would have to be of such a character that they could not serve as the basis of a system for the measurement of time in order to preclude one possibility of a principle for the cross-identification of times. If one then imagined that the experience of the inhabitants of that world were much like the character of the experiences of Icabod and Isabel one might succeed in describing a world the inhabitants of which would be justified in treating it as a world with unitary space and non-unitary time.[13] I will not seek to elaborate an account of such a fantasy world which would be even more outlandish than those described in this chapter. For my essential point — that time need not be unified — has been established by the argument of section 4.

V

THE TOPOLOGY OF TIME III: THE BEGINNING OF TIME

infinity lies in the nature of time, it isn't the extension it happens to have.
Wittgenstein, 1975, p. 164

1 BEGINNINGS AND ENDINGS

Could time have had a beginning? At first glance and perhaps even at second glance posing this question seems to set us on the well-travelled road to antinomy. For instance, if we suppose that time had a beginning, our normal linguistic habits lead us, seemingly inexorably, to talk inconsistently of time before that beginning. To suppose, on the other hand, that time could not have had a beginning will lead us to conclusions which while consistent are unpalatable. For given that change might have had a beginning we are committed to thinking of any world with a first event as a world with endless eons of empty time. And our reluctance to embrace empty time leads us to be chary of admitting this emptiness. For as we will see in section 4 of this chapter, the posit of time before a first event cannot play the explanatory role which temporal vacua play in the fantasy world of chapter II.

There are substantial issues at stake here. Before we can focus on them we need to introduce some routine clarification of the notion of a beginning of time. The notions of beginning and ending are most frequently applied to things in time, and it might seem that in applying these notions to time itself we are in danger of entering a morass of nonsense. Indeed it has seemed to some writers that to pose the question of whether or not time had a beginning is to commit a category error. On this view it is only of things in time that one can legitimately ask of their beginnings and endings. For example, that it begins to snow entails that at one time there is no snow falling and that at a later time there is snow falling. On this construal, to say that time began involves making either the contradictory assertion that, at one time, time did not exist, and then at a later time, it did exist; or the absurd supposition that some super-time exists relative to which time itself can be said to begin. But in point of fact the only moral to be drawn here is the entirely insubstantial one that we cannot think of the beginning of time on an exact parallel with the beginnings of things in time. What we should have in mind when we say that time began is something akin to what we have in mind when, for instance, we say that the natural number series begins with zero. This means that there is no natural number coming before zero when the numbers are taken in the standard order. It is on this model that I will understand the notion of beginning time. That is, the claim that time began is to be understood as the claim that the set of temporal items has a first member under the basic temporal ordering relation. If we take the set of temporal items as the set of instants and the basic relation as that of being temporally before, the hypothesis that time had a beginning is that hypothesis that there is an instant such that no other instant is before it. And the hypothesis that time had no beginning is the hypothesis that for every instant there is a distinct earlier instant.

2 ARISTOTLE, SWINBURNE AND TENSES

The thesis that time could not have had a beginning has been defended by appeal to what are alleged to be logical truths involving tenses. In his recent book, *Space and Time*, Richard Swinburne has argued as follows:

> Time . . . is of logical necessity unbounded. Before every period of time which has at some instant a beginning, there must be another period of time, and so before any instant another instant. For either there were swans somewhere prior to a period *T* or there were not. In either case there must have been a period prior to *T* during which there were or were not swans.[1]

In what follows I will make use of the notion of a *tense-logical truth* which was introduced on p. 63. By a tense-logical truth of such and such form I mean that any proposition of that form would be true at any time in any possible world. I will also make use of some simple tense-logical notation, write 'P_{-}' for the past-tense operator 'it was the case that', and 'F_{-}' for the future-tense operator 'it will be the case that'. I shall take it that in our semantics, simple tensed propositions are assigned the value T (truth) or F (false) at each time and that the complex proposition 'Ps' is assigned T at some time if and only if there is some earlier time at which 's' is assigned T. 'Fs' is assigned T at some time if and only if there is a later time at which 's' is assigned T. In these terms we can clearly distinguish between the propositional forms '$P-s$' and '$-Ps$', the former being the past-tense assertion of the denial of 's', and the latter being the denial of the past-tense assertion of 's'. For '$P-s$' can be true at time t only if there were prior times at which s is false. But '$-Ps$' can be true at time t if either there are prior times and at all those times s is false; or there simply were not times prior to time t.

Consider now Swinburne's premise: *there were swans or there were no swans*. The problem here is that the English sentence 'there were no swans' is *Janus-faced*. It can be used either to deny the past-tense proposition 'there were swans' which we can write as '$-Ps$'; or to assert that past-tense version of the denial of 'there are swans', which we can write as '$P-s$'. Taking this disjunct of the premise in the latter way renders the argument *question-begging*; taking it in the former way does not give us a valid argument at all. Suppose we construe the premises as '$Ps \lor P-s$'. Granting the truth of this, at some time t, allows us to argue from either disjunct to the existence of time before t. But why should we accept this premise as having the status of being tense-logically true? The reductionist will argue that this cannot be a tense-logical truth on the grounds that if there is a first event that proposition is false at the start of that event. Thus, unless some independent argument is advanced for thinking that time of logical necessity had no beginning, we have no grounds for regarding this as being a tense-logical truth and hence the argument is question-begging.

That Swinburne offers no reason for accepting the crucial premise suggests he is taking it in the other formation: namely, as '$Ps \lor -Ps$'. In this case, as we have a substitution instance of the law of the excluded middle we (or at least non-intuitionists) might be expected to accept it as a tense-logical truth. However, we cannot now argue from the disjunct '$-Ps$' to the existence of prior times. For on the understanding of the tense-operators that Swinburne needs for his argument

'*–Ps*' *might be true in virtue of the absence of prior times*. It is only by adding as an additional premise: '*–Ps → P–s*' that the argument goes through. But this additional premise is logically equivalent to our other formulation and for that reason renders the argument again question-begging.

Thus, on neither interpretation does Swinburne's argument go through. The *appearance* of an argument is generated by the fact that in normal contexts we do allow an inference from '*–Ps*' and '*P–s*', *for the obvious reason that we normally operate under the implicit assumption that there were prior times*. Under that assumption, the inference is innocuous. If we wish to represent this standard inference pattern in such a way as to avoid an implicit commitment to non-beginning time we can use, say: '*–Ps → (P(s → s) → P–s)*'. By putting any tautology in the second antecedent we block the inference from '*–Ps*' to '*P–s*' at *t* except under the assumption that *something* was true; i.e., that there was time before *t*.

Swinburne's argument is rather reminiscent of the following argument of Aristotle's for unbounded past and future time:

> Now since time cannot exist and is unthinkable apart from the moment, and the moment is a kind of middle-point, uniting as it does in itself both a beginning and an end, a beginning of future time and an end of past time, it follows that there must always be time. . . .[2]

This amounts to the claim that it is a necessary condition of a moment's ever being present that there are at that time past and future moments. This claim can be represented by the following pair of tense-logical postulates:

(i) $q → FPq$;
(ii) $q → PFq$.

These postulates express the thought that for any present-tense proposition which is true at a time there is then a past time during which the proposition was going to be true and a future time at which it will have been true. Together these postulates, which have commended themselves to others,[3] entail Aristotle's thesis that there is no present that is not, so to speak, flanked by a past and a future. It follows from these postulates that time is non-ending and non-beginning. However, we cannot establish that time is thereby of logical necessity unbounded without first establishing that these postulates are tense-logical truths. But establishing that requires establishing just what is at issue.

One certainly does feel that there is some non-contingent link between a proposition's now being true (or an event's now being present) and its going to be that it was true (or a present event's being past in the future) and its having been going to be true (or a present event's having been future in the past). This feeling gives initial plausibility to Aristotle's argument. One way we can do justice to this feeling without thereby committing ourselves to necessarily unbounded time is to adopt the following technique and take as tense-logical truths the following pair of weaker postulates:

  (i) $q \rightarrow (P(s \rightarrow s) \rightarrow PFq)$;
  (ii) $q \rightarrow (F(s \rightarrow s) \rightarrow FPq)$.

Given these postulates one can pass from p's being true to its going to be that p was true only in conjunction with the assumption that there is a future. This additional condition is captured by applying the future-tense operator to any tautology. For if there is a future, '$F(s \rightarrow s)$' is bound to be true, and if there is no future, it is bound to be false. In Aristotle's idiom we have advanced the weaker claim that nothing excepting a first or a last moment of time can be present unless it is a mean between the past and the future.

One can characterize the style of argument we have been considering as follows. We are offered a sentence schema involving tense operators — a schema that *seems* to reflect an inference pattern we standardly employ. The sentence schema is taken to be what I called a tense-logical truth and it is concluded that time has to have whatever topological properties the schema's being a tense-logical truth would entail. These arguments seem plausible as one may think that grasping the inference patterns represented by the schema is part of what it is to grasp the sense of the tense operators. Swinburne, for instance, speaks of the ordinary concept of time. One imagines that he has in mind the thought that our ordinary concept of time involves concepts of the past, present and future the content of which is in part represented by the inference patterns in question. However, we can represent the rules governing the usual tense operators in ways that will do justice to our usual inference habits without implicitly attributing a particular topological structure to time. As we saw, one might equally offer the following schema: $p \rightarrow (F(p \rightarrow p) \rightarrow FPp)$, $p \rightarrow (P(p \rightarrow p) \rightarrow PFp)$. The general moral that is to be drawn is the following. We cannot establish that time of logical necessity has a given topological property by appeal to allegedly tense-logical truths. For any such argument will need to be supported by a proof that such allegedly tense-logical truths are indeed tense-logical truths and this in turn will require a proof that time of logical necessity has the topological property in question.

3 KANT, INDEFINITE EXTRAPOLATION AND POSSIBILITY

In the Antinomies Kant purports both to derive a contradiction from the assumption that the universe had a beginning in time and to derive a contradiction from the assumption that the universe had no beginning in time. It is not always noted that in both cases the argument takes place under the assumption that time had no beginning. Kant appears to take it that the following consideration adduced in the Aesthetic establishes the non-beginning of time:

> The infinitude of time signifies nothing more than that every determinate magnitude of time is possible only through the limitations of one single time that underlies it. The original representation, time, must therefore be given as unlimited.[4]

If I understand this argument, and I am not at all sure I do, it may come to this. Any particular duration (determinate magnitude of time) is the duration containing such and such instants and not containing all other instants (the limitations of one single time). If we add the implicit premise that for any duration there is a larger possible duration, Kant seems to be claiming that time must be thought of as infinite in order to accommodate the unending sequence of longer and longer durations.

Given a particular metrication of time, if for every event or process which has occurred, some event or process of greater duration had occurred, we should be committed to thinking of time as having infinite past duration, *under that metrication*. However, if we came to think that there was a first event and if our metrication assigned the interval of time lapsed since that event some finite value, we would not be committed thereby to thinking of time itself as pre-existing, so to speak, this first event. We can assign a well-defined finite value to the duration from the beginning of the first event until now without presupposing time prior to that event. To put the point colourfully — time must be thought of as at least as big as any process or sequence of processes in time. But we need not think of it as actually bigger than any actual sequence of processes that occur.

The particularly unconvincing character of the arguments of Swinburne, Aristotle and Kant tempts me to offer the following *ad hominem* explanation of what brings many to think that there must be some cogent argument for non-beginning time. Our uncritical thinking about time is influenced by the technical devices we use for representing the temporal aspects of things. We often label moments at which events occur by elements of some number system. Some auspicious event like the birth of a prophet is assigned, say, zero. Earlier and earlier events are assigned larger and larger negative numbers; later and later events are assigned larger and larger positive numbers. An event taken to be the first is assigned some particular large negative number. As the number sequence itself used can be further extrapolated we are seduced into thinking that there must be times corresponding to these numbers. However, the extrapolation within the representing system does not guarantee that those extrapolated elements of the system have referents. To use an analogy — suppose an interest in a scientifically useful concept of temperature leads us to define temperature in terms of molecular motion and to set $-273.16\,^\circ$C as the temperature of a body whose constituent molecules are not in motion. The mathematical system of the real numbers used to represent the temperature of bodies admits extrapolation below $-273.16\,^\circ$C. But these lower numbers cannot be thought of as designating lower temperatures which as a matter of fact no body possesses. For, by definition, there is no temperature lower than $-273.16\,^\circ$C. I am not here suggesting there is an exact analogy with time. For it is not clear that talk of time before a first event is simply incoherent in the way that talk of lower temperatures than that of a body with no molecular motion would be.

Quinton has recently argued for the infinite extent of time by claiming:

The infinite extent of (Euclidean) space cannot be invoked to support the infinite size of the material world that occupies it in the way that its infinite divisibility was used to establish the infinitely divisible. At however remote a distance we have found matter to be present it is always possible to describe matter further off still. But it does not follow that there is any matter there. As long as there are some material things we can set up a system of spatial characterisations which allows for the description of possible things at any distance whatever. But it is a contingent question whether these positions are occupied, whether these possibilities are realised. Similarly with time. It has no beginning in the sense that there is no date an earlier date than which cannot be significantly described. But there may be a date at no time earlier than which anything was happening at all. Infinity, we might say, is a necessary feature only of systems of description, not merely those which contain numbers but any which contain such transitive asymmetrical relations as 'smaller than' and 'further than' and 'earlier than'. But this entails neither that there is no limit to the minuteness or remoteness of actual things nor, *a fortiori*, that there are infinitely small or infinitely remote things.[5]

Nothing very significant would be established if it could be shown that we are committed to the non-beginning of time by our 'system of descriptions', or by 'the grammar of our determinations of time'.[6] For it only prompts the more substantive question as to whether we should continue to employ that particular system of descriptions or grammar. We may have developed our system of descriptions without taking account of certain possibilities, i.e., that there was or might have been a first event, and the viability of that system of descriptions may depend on that possibility not being realized. In any event Quinton has not established that this is a feature of our system of descriptions. We cannot argue for the non-beginning of time on the grounds that we use a transitive asymmetrical relation in giving temporal order. For transitive asymmetrical relations can be defined on finite sets.

In chapter II we noted a conceptual link between time and the possibility of change. It is not infrequently supposed[7] that one can argue from this conceptual connection to the non-beginning of time along the following lines. Suppose that as a matter of fact there was a first event, E_0, which occurred at time t_0. While E_0 was the first event, it might not have been. It was true that at time t_0 there might have been events occurring at times before t_0. Hence, it is argued, there were times before t_0 at which things might have, but did not, occur. The crucial premise is the claim that there might have been events before t_0. If this is taken to mean that there were times before t_0 at which things might have happened, the argument is question-begging. For what is at issue is whether there were times before t_0. However, the claim is more plausible if construed as follows. In some possible world different from this world, E_0 is preceded by other events. In that possible world there are times before t_0. Hence it is true in the actual world that there *might have been* times before t_0. On this construal, we need some additional premise to license the inference from the possibility of times before E_0 in the actual world to the existence in the actual world of times before E_0.

The above argument turns on the assumption that the possibility of events before a given event is sufficient to establish the actuality of times before the time of the given event. However, it has already been argued in chapter II that neither the logical nor the physical possibility of the occurrence of an event is a sufficient condition for the actual existence of a time at which the event might have occurred. Hence the assumption on which the argument in question turns is not tenable.

In a somewhat similar vein Peirce argued for the non-beginning of time:

> If on a Monday an idea be possible, in the sense of involving no contradiction within itself regardless of all mere circumstances, then it will be possible on Tuesdays, on Wednesdays, and on Fridays; in short it will be possible forever and ever, unless the ideas of the circumstance should come into definite rational contradiction to the idea in question. Consequently, mathematical Time cannot have an arbitrary beginning nor end. For it is a possibleness; and what is possible at all is possible without limit, unless there be some kind of a limit which comes into definite rational contradiction with the idea of Time.[8]

Peirce seems to be arguing that if there was a first instant of time, t, there would be a limitation on logical possibility. For logical possibilities would not obtain before time t. That is so. There was no time before time t and hence no possibilities could obtain before time t. But this result is quite compatible with the principle, appealed to by Pierce, that what is logically possible at one time is logically possible at any other time. For we can hold that at time t it *was* possible that p, if we construe this not as there having been a time at which p might have been true but as p's being the case in some possible world in which there are times prior to t.

4 EMPTY TIME, LEIBNIZ AND EXPLANATION

We have seen that the case for the necessary non-beginning of time is unconvincing. In general those who have agreed have argued by appeal to what I called in chapter II *Aristotle's Principle* that time had a beginning if and only if change had a beginning. If my argument in that chapter is cogent, Aristotle's principle cannot be defended in any strong form that would entail that talk of time before a first event was incoherent or vacuous. However, we can use an argument of Leibniz to support the thesis that it would be inadvisable to posit the existence of time before a first event.[9] For given that the actual world had a first event, one who posits the existence of time before that event has to acknowledge that there is a possible world exactly like this one except for its location in time, and is hence faced with the question: why did the world begin at the time it did begin? One who equates the beginning of time with the beginning of change is not faced with that question. Of course both are faced with the question as to why the world began with the event it did begin with.

It seems *prima facie* that there could be no answer to this question. Consider the usual sort of case where we might ask why some event occurred when it did. Suppose I ask why the Canadian election took place last autumn and not last spring. We might in part explain the autumn election by citing some particular facts (initial conditions) about our political leaders' ability to perceive conditions conducive to their self-preservation, and some general facts (law-like regularities), say, about the inclination of politicians to strive for self-preservation. In part, at least, we explain the election's relative position in the temporal sequence of events constituting the history of our world by citing other things besides the position of the election that would be different in a world in which the election had a different position. But on the assumption that there was time before the first event, there are other possible worlds exactly like this one (assuming it to be a first-event world) except for location in time, and so, *ex hypothesi*, there would be no possibility of explaining why this world has the location it does in time by citing other differences between the worlds.

If it were necessarily true that everything is in principle capable of explanation, we might have an argument to support the thesis that there could not be time before change. As this is not a necessary truth, we do not have such an argument. However, the argument has some force. For it would certainly count against the postulation of time before a first event that it involves postulating a state of affairs (the world's beginning at a certain point in time) which might have been different (it could have begun at another time), but for which no explanation could be given. We have, other things being equal, a preference for not postulating states of affairs for which no explanation can be given. And, perhaps more importantly, we do not see what would count in favour of the postulation of empty time prior to a first event. There are no facts about the world which would be explained by such a postulation. It could only be a sensible postulate with explanatory force in the context of a theory according to which a world without time before its first event would have some other difference besides this difference from a world with time before its first event.

5 ARISTOTLE'S PRINCIPLE, EMPTY TIME AND THE BEGINNING OF THE UNIVERSE

If my earlier discussion (ch. II) succeeded in establishing that there is no incoherence in positing periods of time without change, the reductionist cannot appeal *tout court* to Aristotle's principle, *AP*, in arguing that time before the first event is not a logical possibility. So it would seem that we cannot rule out *a priori* the existence of time before a first event, nor, given the failure of the Platonist arguments, can we assert *a priori* the existence of time before a first event. We are left then with the question of what would be reasonable grounds for asserting the existence of time prior to a first event. For in keeping with our modified version of *AP*, the positing of periods of time without change is sensible only if that posit is part of a theory which fares better than its rivals in giving an account of observable change. It remains possible to argue that no explanatory end would ever be served by the posit of time prior to a first event. An examination of the structure of arguments adduced by cosmologists in support of the contention of a first event seems to support that conclusion. To such an examination I now turn.

I have treated the notion of the beginning of the universe as unproblematic. If there was some incoherence in the very notion of a first event, we should have an argument for the non-beginning of time as a matter of logical necessity *without* a violation of Aristotle's principle. Leaving Kant aside for the moment, the most frequently encountered argument for this conclusion is perhaps the following. It is permissible to talk about the beginning of processes in the world, but to transfer this talk to the world itself is to commit some form of category error.[10] Certainly we must take care in transferring to the world itself notions paradigmatically applied to sub-systems of the world. For example, the notion of entropy raises certain problems of this kind. For if we have defined entropy as a property of a closed system, it is not clear how we can extend this notion of entropy to the entire universe. The notion of a closed system is defined in such a way as to make it unclear what could be meant by considering the universe itself as a closed system. However, the only moral to be drawn here is that care must be taken. The hypothesis that the universe had a beginning is conceptually un-problematic if it is taken as the hypothesis that the set of all past events had a first member.

Kant purported to find a contradiction in the notion of a beginning of the universe:

> For let us assume that it has a beginning. Since the beginning is an existence which is preceded by a time in which the thing is not, there must have been a preceding time in which the world was not, i.e. an empty time. Now no coming to be of a thing is possible in an empty time, because no part of such a time possesses, as compared with any other, a distinguishing condition of existence rather than of non-existence; and this applies whether the thing is supposed to arise of itself or through some other cause. In the world many series of things can, indeed, begin; but the world itself cannot have a begin-ning, and is therefore infinite in respect of past time.[11]

Kant assumes that time could not have had a beginning and is arguing that if there was a first event there could be no explanation as to why it occurred at the time it did occur. Even if it were granted that every-thing must (logically) be capable of an explanation, it does not follow that there is an incoherence in the notion of a beginning of the universe. For one could deny equally the assumption which Kant has failed to justify, that time is of necessity non-beginning. If we take time and the universe as beginning together, so to speak, there is just no question as to why it began at the time it did begin.

More interesting difficulties arise if we consider the epistemological problem of what would constitute evidence that our universe had a beginning. Sceptics about induction are apt to feel particularly sceptical with regard to claims about the state of the universe a very long time ago. And those who are not simply sceptical about induction find their courage falters in the face of the degree of extrapolation involved in cosmological arguments. But if we are not sceptics about induction in general, we have no reason to falter, though we ought to have only a modest confidence in our beliefs concerning the very distant past.

Setting aside these general doubts as not having particular relevance to the questions of the beginning of time and the beginning of the universe, there are, none the less, certain peculiarities in the sort of argument standardly adopted in support of the hypothesis of a beginning of the universe that do give rise to interesting and perhaps intractable epistemological problems. Arguments to support the hypothesis of a beginning of the universe typically have the following general structure. We suppose that we have some evidence that the radius, R, of the universe at time t is given by the following expression:[12]

$$R(t) = (at + b) \qquad\qquad (a, b, \text{ constants}),$$

Density, $\rho(t)$, is then given by:

$$\rho(t) = e\,\frac{(\mathrm{d}R)^2}{(\mathrm{d}t)} \cdot \frac{K_2}{R}$$

In this case there is some value t_0 of t for which $R(t)$ is zero in a model with $b = 0$. Under the assumption that matter has not, since time t_0, been created *ex nihilo*, the universe would have been at time t_0 in a state of infinite density (call this state S_0). We suppose further that we have evidence supporting other laws which are incompatible with the existence of a state S_0 of infinite density. Thus, it is reasoned, S_0 could not have obtained, and the first state of the universe must have been some state, S_+, occurring at some time t_+ at which $R(t_+)$ is small but not so small as to give the universe an impossibly high density.

I am not concerned with the question of evidence for the alleged law-like regularities embodied in the expansion function; nor for the alleged law-like regularities which are incompatible with the existence of infinitely dense states of the universe. Rather we are concerned with the identification of the state S_+ which we assume to have existed as the first state of the universe. Not infrequently we have evidence for a pair of laws which are discovered to have incompatible consequences in some contexts that fall within their scope. It is sometimes fruitful to seek a modification of one or other of the laws (given that they are highly confirmed over some range of data and do not have incompatible consequences over that range of data) which brings them into harmony. In the present case we are invited to correct the expansion function on the basis of the density laws. But this necessitates abandoning another well-entrenched belief, namely the belief that at least at the macroscopic level all states of the universe are causal upshots of temporally antecedent states. In correcting the expansion law, we are forced to admit the existence of an uncaused state, S_+. One might argue the proper conclusion to be that S_+ is a state beyond which we cannot extrapolate on the basis of our present theories. It is more reasonable, the argument would run, to stick by our belief in the causal principle (applied macroscopically) than to stick by our beliefs in the particular expansion function and the particular density laws in question.[13] No matter how much evidence we had for the density laws and the extrapolation laws, it might be more reasonable to appeal to the causal principle and regard S_+ as a singularity whose causal antecedents cannot be described, rather than regard S_+ as a state without causal antecedents.

Even if we came to hold that S_+ was an uncaused state, the assertion that S_+ was the first event depends on further assumptions. To bring this out, suppose we have a theory which involves the above expansion and density laws and which also involves a function giving the mass of the universe as a function of time, such that the derivative of mass with respect to time is negative and constant. Extrapolating backwards we always have non-zero mass; extrapolating forwards we have, for some t, zero mass. If we suppose the metric tensor giving the space–time structure of this world collapses for zero mass, we seem to have reasons for thinking not just that there was an uncaused state S_+ but that there will also be a last state of the universe. This might look like evidence for the hypothesis that this universe had a first state S_+ and a last state S_n. However, the assertion that S_+ is uncaused undercuts any ground we might have had for denying that S_n will be followed by some uncaused event. That is, if we allow the violation of the causal principle we have no grounds for denying the existence of an uncaused state, perhaps rather like S_+, following S_n. Analogously, if we assert that S_n is the last state, we commit ourselves to giving up the principle of conservation of matter-energy, in which case we have undercut the grounds we might have had for denying that S_+ was preceded by state S_0 which, as it were, vanished without trace. In taking the uncaused state S_+ as the first state we are implicitly appealing to the principle that we have jettisoned in talking of the end of the universe.

This interplay between the causal principle and the principle of conservation of matter-energy suggests that we will never be in a position to justifiably assert the existence of both a beginning and an ending of the universe. And, in relation to the earlier argument concerning a beginning universe, it brings out that we are implicitly appealing to a principle of conservation in passing from the claim that S_+ is uncaused to the claim that S_+ was the first state of the universe, the argument here being that any previous states would satisfy the principle of conservation and hence would not vanish without trace.

It is thus more reasonable to plead ignorance about the causal antecedents of some very dense state of the universe a long time ago than to assert that the state was an uncaused first state. However, if one is led to assert the existence of a first state on the basis of beliefs in the appropriate expansion, density and conservation laws, we should deny the existence of earlier times. For if there are earlier times, there is clearly a violation of the conservation principle. For a state with zero matter-energy would be followed by a state with non-zero matter-energy. Thus the line of argument which assumes the non-violation of a conservation principle must take the time of the first state as the first time.

For reasons of the above sort it is difficult to envisage within our current scientific framework any viable theory that involves positing both a first event and time before that event. If such a postulation is not to be entirely idle, the theory would have to involve the claim that there must be some difference between a first-event first-time world and a first-event, no first-time world (other than that difference) — a difference which the theory would account for by reference to time before the first event. This is difficult to envisage. In any event, it would clearly involve violations of the principle of the conservation of energy. Thus, it could not, for instance, come out of the field equations of the General Theory of Relativity as they have built-in conservation principles. And, as we remarked above, once one allows massive violations of the conservation of energy, one loses one of our most important methodological tools used in guiding our theory construction. Certainly one might envisage minor systematic violation of the principle of the conservation of energy. However, what would be involved in a first-event world with empty time prior to that event would be massive violations of conservation of energy. Hence the postulation of empty time prior to first event is idle. Leibniz's remark that 'time without things, is nothing else but a mere idle possibility' *may* indicate that he held this view concerning time before a first event (i.e., that while there is no incoherence involved in talk of time before a first event, there would be no point in positing the existence of such time).[14]

The argument of the particular form that has been considered depends on the assumption that states of infinite density are physically impossible. This is, in fact, a contentious assumption. For it has become common to interpret the work on singularities in General Relativity as showing that that the theory admits of the possibility of 'black holes' which are singularities where density is infinite.[15] In which case we are no longer licensed to postulate the state S_+ as the first state on the grounds that any state prior to S_+ would have a physically impossible density. It appears that the prospects for ever having evidence for a genuine first event are remote. For, supposing that the Big Bang emerged from a singularity of infinite density, it is hard to see what would constitute a reason for denying that that singularity itself emerge from some prior cosmological goings-on. And as we have reasons for supposing that macroscopic events have causal origins, we have reason to suppose that some prior state of the universe led to the production of this particular singularity. So the prospects for ever being warranted in positing a beginning of time are dim.

VI

THE TOPOLOGY OF TIME IV: THE MICRO-ASPECTS

Now 'tis certain we have an idea of extension; for otherwise why do we talk and reason concerning it? 'Tis likewise certain that this idea, as conceiv'd by the imagination, tho' divisible into parts or inferior ideas, is not infinitely divisible, nor consists of an infinite number of parts: For that exceeds the comprehension of our limited capacities. Hume, 1960, p. 32

1 THE MICRO-STRUCTURE OF TIME

In this chapter I take up the question of how we should think of the 'small-scale' aspects of time's topology. Could time be *discrete*? That is, could each instant of time have a uniquely next instant? Or must time be *dense* so that between any two distinct instants there is another instant? A third possibility is that time is not merely dense but is also *continuous*. In explicating the content of these alternative hypotheses, we will start with a brief account of the mathematical notions of discreteness, density and continuity.

The natural numbers are the infinite sequence of counting numbers 0, 1, 2, ... This sequence is discrete, which means that for any member of the sequence there is a unique next number. More generally, any set S ordered by a relation $<$ is discrete if and only if for each member x of S there is a unique member x' of S such that $x < x'$ and if $x < y < x'$ then $y = x$ or $y = x'$. Another example of a discrete order is the set of all positive and negative integers . . . $-2, -1, 0, 1, 2, . . .$ under the standard ordering. In the case of time, let us for the moment assume that there is a set of instants whose structure can be adequately described using the relations of being temporally before and its converse, being temporally after. Under these assumptions, one way of explicating the hypothesis that time is discrete would be to say that for each instant of time there is a unique instant before that instant and a unique instant after that instant. If we have a set S and S' with ordering relations R and R' defined respectively on S and S', it may be that if S and S' have the same number of members we can define a function f which associates with each member of S a unique member of S' which preserves the ordering of S and S'. This means that for a, b in S, we have $R(a, b)$ if and only if $R'(f(a), f(b))$. In this case the sets as ordered have the same structure, a fact which we express by saying that these ordered sets are *isomorphic*. If time is discrete in the sense outlined above the temporally ordered set of instants will be isomorphic to the integers or some sub-set of the integers depending on whether time has a beginning or not and on whether time has an ending or not. For example, if it is discrete and has neither an end nor a beginning, it will be isomorphic to the integers. If it has a beginning but no ending it will be isomorphic to a sub-set of the integers having a first member but no last member.

A set S ordered by a relation R is dense if and only if for any pair of distinct elements a, b in S there is another element c which is such that Rac and Rcb. The rational numbers (the positive and negative fractions) under the standard ordering are dense. If time is dense there will be another instant between any pair of distinct instants. And time will be isomorphic to the rationals or to some sub-set of the rationals depending, as before, on whether time has a beginning or not and on whether time has an ending or not. While the rationals are dense there is a sense in which there are 'gaps' in the rational numbers. There is, for example, no rational number whose square is 2. In order to fill these 'gaps' we add to the rationals the irrationals which are numbers that cannot be represented as fractions. The resulting system is the real number system whose salient characteristic is that it is not only dense but lacks 'gaps' — an idea which we express by saying that the real number system is *continuous*.

We can define the notion of continuity given a set S and an ordering of that set with the help of the notion of a cut. A cut of the set S is a division of S into two sub-sets, S' and S'', such that (1) each member of S is in either S' or S''; (2) no member of S is in both S' and S''; (3) each member of S' comes before in the ordering any member of S''. S is continuous under R if and only if any cut of S into sub-sets S' and S'' is such that either there is a unique least member of S'' or a unique greatest member of S'. The rationals can be seen not to be continuous if we consider the cut of all rationals into the sub-set of all rationals whose squares are less than 2, and the sub-set of all rationals whose squares are greater than 2. There is no unique greatest member of the former and no unique least member of the latter. This definition of a cut for the rationals does not give a cut of the reals. For $\sqrt{2}$ will not be a member of either set. If we take a cut by defining a sub-set of reals whose squares are all less than 2 and a sub-set of reals whose squares are greater than or equal to 2 this latter sub-set has a unique least member, namely $\sqrt{2}$. As any cut of the reals produces sub-sets S' and S'' which are such that either S' has a greatest element or S'' has a least element, the reals are continuous.

If time is continuous the set of all instants under the ordering given by being temporal before will be isomorphic to the reals or to some sub-sequence of the reals. It should be noted that in explicating the notions of discreteness, density and continuity as applied to time, we assumed that the structure of time could be adequately characterized by the two-place relation of being temporally before. Modifications in these explications of a straightforward kind would be required if, for example, we were treating time as closed. In this particular case, as we saw in chapter III, the appropriate ordering relation is the four-place relation of pair-separation.

2 DISCRETE TIME

It is quite standardly argued that there is something conceptually problematic in the idea that time might be discrete in the sense explicated above. For instants are extensionless parts of temporal intervals and if time were discrete any extended period of time would have a finite number of durationless parts of instants. And it is hard to see how adding up a finite number of unextended temporal instants could give an extended temporal magnitude. This feeling has been articulated with regard to space by Nerlich as follows:

> But if an interval has length and the interval divides into grains, as it does in discrete space, then the grains cannot have zero length, for they must simply add up to the length of the interval. It is beyond

me to grasp what concept of length would permit any conclusion other than this simple one.[1]

It may well be that those who assume that time might be discrete are guilty of thinking in terms of the picture of extensionless time-points separated in time in the way in which one might draw a sequence of spatially separated points. In that case, the extension of the interval spanned by the sequence of points is derived from the extended intervals between the points. Clearly this will not do, for it an interval ultimately consists of a set of durationless instants there is nothing between the instants which could generate the extension of the interval.

In part, the problem involved in trying to conceive of time as discrete arises from the fact that we think of extensions (spatial or temporal) as having parts which are themselves extended. That is, we think that any temporal interval has a proper part (a part not equal in duration to the interval) which is itself an interval having a proper part and so on. Under this assumption of the infinite divisibility of extension, any finite duration has an infinite number of finitely extended parts. But if time were discrete in the sense outlined in the last section a finite interval will have only a finite number of extensionless parts and only a finite number of extended parts. Thus our conception of temporal intervals as infinitely divisible leads us to regard time as not being discrete. In view of this conceptually oriented, *prima facie* case against discreteness, it is worth considering what has prompted some to take seriously the possibility of discrete time. The line of reasoning to be considered is unsatisfactory but it is instructive.

It is sometimes said that there is a smallest sub-atomic particle, its diameter being referred to as a *hodon* of length. Given that the velocity of light is the fastest possible velocity we can calculate the time that the fastest velocity could traverse the smallest diameter. The resultant value is called the *chronon* of time. We are told[2] that if the theories leading to these calculations are correct, the chronon is the smallest interval of time. This is sometimes represented as conflicting with the claim that time is infinitely divisible. However, the chronon is an extent of time and it has parts which are extents of time. No matter how important these measures, hodons and chronons, should become in some future scientific theory, this remains true. For a light wave propagates across the hodon in a chronon and where the light wave is part way across the hodon of length, part of a chronon of time has passed. And part way through a chronon of time the light front has got part way across the hodon of length. Thus the theory which is employed in introducing the notion of a chronon and a hodon presupposes that it is legitimate to refer to temporal intervals smaller than a chronon. The hodon and chronon ideas are developed in the context of a theory presupposing at least dense if not continuous change. For light does not jump across the spatial intervals of a hodon, it continuously traverses the spatial distance.

Even if time is infinitely divisible it is possible that we might discover there to be a sort of minimum threshold of *operationally specifiable extents of space and time*. Consider some extended spatial object, say a table. We can sensibly talk about the first $1/n$th of the table from the corner of the table. If we have some object that fits, under the appropriate operation, exactly n times along the edge of the table, we can identify the part in question as the part that would be marked out by this object if appropriately placed at the edge of the table. If in fact we do this, I shall say that we have *object-marked* the first $1/n$th part. If there are smallest objects, there will be some n such that the $1/n$th part cannot be object-marked with the given objects. Even under this assumption we can talk sensibly of still smaller parts. The $1/n$th first part, where n is larger than the number of times the smallest object fits along the table, means that part that would be marked by an object which did fit n times along the table.

In a similar manner, given some interval of time, say the interval marked by the last rotation of the hour hand of my clock, we can talk of the part of that interval marked by, say, the movement of the hour hand from the 1 to the 2 on the dial. In this case we might say that this part of the interval is an *event-marked* part. I can still sensibly refer to the first $1/n$th part of the duration even if I cannot describe some event which took just that part. I mean by the first nth part, the part that would be marked by the first occurrence of some repeatable event whose repetitions are self-congruent where n successive occurrences of that event would be equal in duration to the interval in question. It will not be possible in the case of the discrete quantized change world of chapter II to event-mark the parts of the mini-freeze. Any spatial extent that cannot be object-marked as a matter of fact, and any duration that cannot be event-marked as a matter of fact, will be said to be lengths and durations that are not operationally specifiable.

If there are objects of smallest size, those objects set limits to operationally determining spatial parts. Given also a maximum velocity, this sets operational limits on the determination of temporal parts of temporal durations. If we deem spatial lengths less than a hodon not operationally specifiable, we will deem temporal durations less than a chronon as not being operationally specifiable. For, say, half a chronon cannot be specified operationally as the interval taken by light to cross half the hodon for the half-hodon distance cannot be operationally specified. Thus, while there might be, in this sense, limits to how finely we can operationally specify temporal intervals or spatial distances, this does not mean that there are not smaller distances and shorter times. It means only that *certain* ways of referring to these smaller bits of space and time are not available to us. The smaller bits might still be there to be referred to if we choose to.

If it is legitimate to refer to ever smaller parts of durations even though those parts may not be operationally specifiable, we are constrained to regard time as infinitely divisible and hence as not being discrete. Someone with an operationalist orientation to measurement might well deny the legitimacy of talk of durations smaller than any which can be event-marked. It could then be argued that time is not infinitely divisible as there are smallest durations, and hence that time is discrete in the sense that any finite interval of time has among its parts a discrete set of these minimal parts adding up to the interval in question. Nerlich,[3] in his discussion of discrete space (a discussion which can be carried over with equal cogency to time), understands the thesis of discrete space as the thesis that the process of division of spatial intervals into parts terminates in indivisible but extended 'grains'. However, given the anti-operationalist, anti-verificationist theme that runs through Nerlich's book in which this discussion of discreteness is to be found, it is surprising that he takes the notion of indivisible grains as unproblematic. For without some fairly strong operationalist/verificationist assumptions, there is no reason not to argue that any 'grain' in virtue of being extended must have extended parts.

Is this the end of the matter? Are we to conclude that having rejected the required strong form of operationalism/verificationism, we are committed to the thesis that time is infinitely divisible and hence not discrete? I think not. At best the argument shows that our concept of extension as presently constituted commits us to infinite divisibility, in so far as we accept that concept as legitimate for application to the world. The argument from divisibility does not purport to do more than to establish the density of time. However, we standardly represent time as not merely dense but as also continuous. While attempts are frequently made to establish density by *a priori* means, it is only rarely that such *a priori* arguments are adduced to support the claim that time is continuous. One notable exception to this claim will be critically evaluated at the end of this section. Why then do we tend to regard time as not merely dense but also continuous? Setting aside until the next section of this chapter the question as to whether we *should* so regard time, the answer is quite simply that the best physical theories we have in fact constructed of the physical world require in their mathematical formulation a time parameter that ranges over the elements of the real number system. This means that we map the interval of time marked by some event onto intervals of real numbers, and the intervals marked by the extended parts of the event onto non-empty sub-sets of the interval in question. We take individual real numbers as denoting instants and project back onto the interval a non-denumerably infinite number of extensionless parts or instants, one corresponding to each real number in the interval. That is, our belief in the continuity of time does not arise from any argument relating to infinite divisibility, it arises from our projecting onto the world the richness that is present in the mathematical system which we have found to date to be essential to the construction of viable physical theories.

When it is seen that this is the source of our conception of time as continuous, we can see what might lead us to adopt a conception of time as discrete. For suppose it should transpire that the best scientific theories of the physical world represented time by a parameter which ranged not over the rationals, nor over the reals, but over the integers. The interval of time marked by a finite event would be assigned a finite interval of natural numbers. Any extended proper part of that interval would be assigned a sub-interval of the interval in question. As before, the individual integers would be thought of as designating the extensionless parts of the interval. If this were our device for representing time, we could not represent any interval of time marked by a finite event as having an infinite number of finite parts. It would have a finite number of extensionless parts. Given that we project continuity onto the world because currently our best theories involve representing time by a continuous time variable, if our best theories should in the end turn out to involve representing time by a discrete time variable, we would have at least as much reason to regard time as discrete as we now have for regarding it as continuous.

If these speculations came to represent the best received scientific view, we should regard time as not infinitely divisible. In this case it would not be because time intervals had a finite number of indivisible but extended parts (an assumption which we have found to be problematic). It would be indivisible because each finite interval had only a finite number of indivisible grains which would be indivisible in virtue of being unextended. Certainly it seems counter-intuitive to think that a finite number of extensionless grains should 'add up' to an extended interval. But at the level of intuition it is equally puzzling how even a non-denumerably infinite number of extensionless grains could 'add up' to an extended interval. It took literally centuries of efforts to school our intuitions before this result came to seem (to some at least) as intuitively plausible. Similarly, should it turn out that the best physics is based on the representation of time by a discrete variable, we ought to further school our intuitions. Perhaps we can do something to make the result seem less problematic. For reasons to be given in section 8 of this chapter intervals should be taken as basic. Undoubtedly there are intervals. If I move my hand from left to right, I mark out an interval of space and this event marks an interval of time. If I then move my hand back to the left I mark out a second interval of time. And we find no difficulty in understanding how one extended interval of time can be added to another to produce a greater extended interval. The investigation of the micro-structure of such intervals is a highly theoretical affair and is to be discovered through an examination of the character of our best physical theories. We must be wary of carrying over intuitions about extended intervals to our thoughts about the extensionless parts of such intervals. If we do so, we will be led to suppose that you cannot get an extension greater than that of a or b by taking a and b together unless a and b have extension themselves. This is where our intuitions need schooling and it may help to introduce an analogy. We are familiar with the fact that an extended coloured object will have very small parts which are not themselves coloured. A certain arrangement of a certain type of particles gives an object a property, colour, that its parts lack. An assemblage of successive instants which lack duration will have a property, namely extension, that each of these particular parts lack.

One attractive feature of discrete time is that it admits of the development in theory at least of a particularly simple measure for the length of intervals. Let a, b, c . . . be variables ranging over instants in discrete time. We will write (a,b) for the interval of time having as its extensionless parts the instants a, b and all instants between a and b. We assign a duration of 0 to each instant and we assign as a measure of duration of the interval (a,b) the number of instants which are part of that interval. Any measure of distance, spatial or temporal, must satisfy the following axioms:

1. $m(a,b) \geqslant 0$
2. $m(a,a) = 0$
3. If $a \neq b$, $m(a,b) > 0$
4. $m(a,b) + m(b,c) \geqslant m(a,c)$.

It is obvious that the measure defined above does satisfy these conditions. The caveat 'in theory at least' is intended to draw attention to the fact that even if it should turn out that it is best to treat the time of this world as discrete, we will obviously not actually measure the length of an interval by counting up its extensionless parts. We will continue to use the standard array of time-keeping devices, results of which would be regarded as more or less approximating to the results of this theoretical measure of duration. It should be noted that this measure will not satisfy the conditions of a measure in the sense of measure theory. For a measure in measure theory must be countably additive. That is, the measure of the union of any set of sets must equal the sum of the measure of the sets forming the union. The set consisting of the union of all singleton sets of points of a given interval will not satisfy this condition. For the measure of each singleton set will be zero and hence the sum of the measure will be zero but the measure of the union of these sets will be the measure of the interval which is not zero. That the measure introduced does not satisfy this condition does not seem to matter. For measure theory is intended to cover measure defined on *all* sub-sets of a given set. But in so far as we are concerned with time, we should only be concerned with the measure of instants and intervals, and measures defined on all sub-sets of a given set need not concern us.

We have found the argument that time must be dense to be unconvincing. While it has been quite common to argue to the contrary, it is only rarely that one finds attempts to establish *a priori* not only density but continuity. One of these rare exceptions is found in Poincaré's argument purporting to show that space must be continuous.[4] As there are good reasons for thinking that space and time should be regarded as having the same order type,[5] his argument, if cogent, would provide grounds for regarding time as continuous. Poincaré's claim is that if space is merely dense the diagonal of a square the sides of which are tangents to an inscribed circle would not intersect the circle at a point. While this is so, it does not reveal anything paradoxical. It only shows that the notion of intersection needs modification. This can be achieved if we understand the intersection of the diagonal by the circle as the

the existence both of parts of the diagonal inside the circumference of the cirle and of parts outside the circumference of the circle.

3 THE EMPIRICAL SIGNIFICANCE OF CONTINUITY POSTULATES

While time can coherently be supposed to be discrete we have no good reasons for taking seriously the hypothesis that it is so. For no one has been able to produce viable physical theories that treat time as discrete. Indeed, all mainline physical theories represent time by a parameter ranging over the real numbers and in so doing treat time as continuous. Interestingly, we will see in this chapter that we can construct equally viable alternatives to these physical theories in which time is treated as merely dense and not continuous. Thus, we cannot dismiss the claim that time is merely dense in the way we dismissed the hypothesis that time is discrete. In effect, we dismissed this latter hypothesis because we have no reason at present for believing in the truth of theories that treat time as discrete. However, as we will see, whatever constitutes a reason for believing in the truth of theories that treat time as continuous will constitute equally a reason for believing in the truth of counterpart theories that treat time as merely dense. Thus we cannot defend the treatment of time as continuous on the grounds that our only viable theories require such a treatment: equally viable theories can be constructed that do not require treating time as continuous.

In any precise formulation of Newtonian mechanics[6] there are postulates that represent space and time as continuous. In what follows, I will show how to construct an equally viable alternative to Newtonian mechanics in which space and time are treated as merely dense. In an analogous manner, the technique employed in constructing this example can be extended to generate rival counterpart theories assuming continuity in which continuity postulates are dropped in favour of mere density postulates.

When we measure directly the value of any parameter such as, for example, length, weight, duration, our technique of measuring will have some limit of accuracy. Consequently, we speak of any given measuring technique as being accurate up to some finite number of decimal places. As any terminating decimal can be represented by a fraction, we need use only the rational numbers in giving the measured value of any parameter. There is no way in which one could come to record an actual measurement by means of an irrational number. For an irrational number cannot be represented by a terminating decimal. However, one can suppose that the parameter really has an irrational value that the measured, rational, value approximates. There would, however, be no means by which such a supposition could be verified or falsified by direct measurement.

Newtonian mechanics is standardly formulated through the use of differential equations. The notion of a differential of a function is only well defined for continuous functions, that is, functions defined over sets having the order-type of the real number line. The differential of a continuous function $x(t)$ with respect to time is defined as follows:

$$\frac{dx(t)}{dt} = \lim_{h \to 0} \frac{\Delta x(t)}{h}$$

We can define an analogue of differential equations for functions defined on sets that are only dense but not continuous, that is, for functions defined over sets having the order-type of the rational number system. To this end the notion of a difference equation is defined as follows:

$$\frac{\Delta x(t)}{h} = \frac{x(t + h) - x(t)}{h}$$

The second difference of $\Delta^2 x(t)$ with respect to h is given by:

$$\frac{\Delta^2 x(t)}{h^2} = \frac{x(t + 2h) - 2x(t + h) + x(t)}{h^2}$$

A difference quotient gives the *average* rate of change over the interval h where a differential gives the *instantaneous* rate of change at a point. The theory of difference equations can be developed in a fashion that parallels the development of the theory of differential equations. For any differential equation on a continuous set there is a difference equation on the appropriate corresponding dense set that approximates the differential equation as accurately as one likes.

Consider Newtonian mechanics. In a rigorous formulation of Newtonian mechanics one postulates that space and time are continuous. Time is represented by a parameter ranging over the real numbers and space by a parameter ranging over triples of real numbers. The crucial laws of Newtonian mechanics are represented by a family of differential equations. For instance, the familiar law relating force, mass and acceleration is given as follows:

$$F = m \, \frac{\mathrm{d}^2x(t)}{\mathrm{d}t^2}$$

In principle, if not in practice, there is a rival theory to Newtonian mechanics which I will call *Notwen's mechanics*. Notwen, unlike Newton, was a perverse fellow corrupted by philosophy who postulated that space and time are merely dense and not continuous. Consequently, Notwen represented time by a parameter ranging over the rational numbers, and space by a parameter ranging over triples of rational numbers. Notwen had the same general ideas as Newton concerning the interrelation of force, matter and acceleration. However, operating with his more parsimonious ontology he could not avail himself of differential equations and was consequently led to slightly different laws which were represented by a family of difference equations. For example, his force law is the following:

$$F = m \frac{x(t + 2h) - 2x(t + h) + x(t)}{h^2}$$

Here h represents a rational interval. This equation can be made to approximate $F = m \, \dfrac{\mathrm{d}^2x(t)}{\mathrm{d}t^2}$ as closely as one likes by taking h to be a sufficiently small interval. In effect, Notwen's mechanics deals with average velocities and accelerations rather than instantaneous velocities and accelerations. The h in the equation above represents a rational interval over which the averages are being taken. By making h sufficiently small, Notwen's equations can be made to approximate the Newtonian ones as closely as one likes. It is important that in specifying his theory Notwen does not specify a particular value for h. Rather his claim is that there is some rational value h for which the theory will fit all the data.

Notwen's and Newton's theories are clearly incompatible. For Notwen claims that time is dense but not continuous and Newton maintains that time is both dense and continuous. However, the theories are, in the appropriate sense, empirically equivalent. Notwen and Newton will test their respective theories by measuring the values of certain parameters which will then be plugged into the equations in order to predict the value of some other parameter. This latter parameter is then checked against the prediction. As we noted previously, the measured values with which they both begin will be represented by rational numbers. Notwen's theory will lead him to predict that the other parameter has some

123

rational value — a prediction which we are supposing subsequent measurement bears out. While Newton's theory may lead him to predict that the parameter in fact has some irrational value he will never be able to confirm that it has an irrational value. For on actually measuring that parameter, Newton will come up with a rational value. If the measured rational value is sufficiently close to the predicted irrational value, he will regard this as evidence for his theory. Roughly speaking, Notwen's theory predicts the parameter to have the rational value it is subsequently measured as having; Newton's theory may predict the parameter to have an irrational value which the subsequent measured rational value is regarded as approximating. Consequently, there will not be (even in principle) any measurements that could be made that would support Notwen over Newton or Newton over Notwen. While the theories in question are incompatible, all the relevant data that could be gathered will leave the choice between the theories undetermined. Any observation that confirms one confirms the other and any observation that falsifies one falsifies the other.

To have an adequate theory of mechanics one needs to do more than develop a system of difference equations which mirror the empirical predictions of the system of differential equations which constitute the core of Newtonian mechanics. One needs in addition, for instance, to assume a geometry for space. Notwen cannot avail himself of a differential geometry such as Euclidean geometry for that involves continuity assumptions. From Notwen's point of view, it would be nice if one could develop a difference geometry analogous to Euclidean geometry. There are, however, certain technical problems involved in this project. For, as we noted in the last section, in *standard* measure it is not possible to give non-trivial metrics for denumerably infinite sets. And under the density assumptions there will only be a demerably infinite set of temporal instants and spatial points. I am inclined to regard this as a mathematical problem which it would be interesting to attempt to solve. However, even if this problem should prove intractable, it is not a decisive objection to Notwen. For Notwen could use the full range of mathematical techniques employed by Newton. In this case, in using an interval of the real number line in representing, for instance, the instants of time in an interval of time, Notwen would regard the non-rational reals as specifying ideal elements added for heuristic purposes and not as specifying actual instants of time. Only the rational reals would be regarded as identifying instants of time. That is, Notwen only affirms the existence of some of the items talked about in this theory. Talk of the other items is a convenient fictional device.

It may be objected that I have not established that no empirical discovery could decide between these theories on the grounds that the history of physics reveals the relevance of certain wider, more general empirical grounds than I have considered for holding one theory to have greater verisimilitude than a rival theory. To see how this objection might be developed, consider a situation in which the actual available observational data underdetermines the choice between two rival theories. It might be that there is a more general theory of wider scope in which only one of these theories can be embedded. Given that the wider theory has some degree of empirical success, it would be reasonable to opt for the theory compatible with that theory. Zahar has recently argued[7] that in 1905 the available data underdetermined the choice between Special Relativity and the Lorentz weather drift theory, but that only the Special Theory was or could be embedded in a gravitational theory (the General Theory of Relativity), and that this constitutes the empirical grounds for preferring Einstein to Lorentz. This situation might well arise in relation to the closed time/open time example of chapter III. For it might be that only one of the two rival theories is compatible with the best total physical theory we can devise where that theory as it turns out does not have, as far as we can tell, an empirically equivalent rival. While this is a possibility, there is no reason to assume *a priori* that the best total physical theory (if there be such a theory) will decide between the rival hypotheses, or that there is a unique best total theory as opposed to two empirically equivalent rival total theories one of which favours the closed time hypothesis and the other of which favours the open time hypothesis. Consequently, I am inclined to concede that I have not established this example of underdetermination. However, it does serve to establish that there is no reason to assume that such a situation cannot arise. Thus, the ball is put back into the court of one who insists that underdetermination is not possible.

In any event this style of objection does not seem to have force against the dense time/continuous time example. For it would seem that a more general theory will decide in favour of, say, continuous time only if it involves some continuity postulates or other. And, using the devices I employed, one could construct an empirical equivalent rival of that theory which employed mere density assumptions. This theory would then decide in favour of the dense but not continuous space and time. So regress to a more general theory in the case of this example will leave the example as a genuine case of underdetermination.

How ought we to respond to this undecidability result? At least two responses merit serious consideration. First, one might simply conclude that the argument shows a limit to the possible extent of human knowledge. Either the world is such that it is true that time is continuous or the world is such that it is true that time is merely dense. Assuming we have been able to reject any other possibility (i.e., that time is discrete), we will remain in ignorance as to which of these two possibilities obtains. I will call this the *Ignorance response*. For it involves assuming that there is some matter of fact at stake here, a matter of fact about which evidence is just not to be had. Alternatively there is what we might call the *Arrogance response*. If we can form no conception of what would constitute evidence for thinking that the facts about the world made one of these hypotheses true or likely to be true rather than the other, we are not entitled to assume that there is some matter of fact at stake here. The claim here is that if there is a matter of fact at stake we can, in principle at least, come to have evidence concerning the facts, if we cannot come to have evidence concerning the truth if there is not a matter of fact at stake. On this response the sentences 'Time is continuous' and 'Time is dense' are not seen as having a meaning which renders them capable of being used to make conjectures about the facts. They are not thought of as being, strictly speaking, true or false. Whatever role they play, it is not the role of the putatively fact-stating proposition. These responses will be further elaborated with a view to adjudicating between them in chapter X.

4 THE POINTS OF TIME

In non-philosophical moments we are quite happy to assume unproblematically the existence both of intervals and of instants of time. We take it to be appropriate in the physical sciences to allow our time parameter to range over intervals of real numbers, each interval of real numbers being taken as designating an interval of time and each individual real number being taken as designating an instant. When in our philosophical moments we ask ourselves what these things are which we have in other contexts been talking happily about, it is instants that seem most problematic (sufficiently so that to say that time is composed of durationless entities called 'instants' becomes an explanation of the obscure in terms of the even more obscure). Consequently we expect any viable theory of time to provide an account of what temporal items such as instants and durations are. It is these questions which we will seek to answer in this and the following sections of this chapter.

Instants contrast with durations. Durations are periods or intervals of time, instants are durationless items a non-finite number of which are parts of any duration. But remarks of this sort do not tell us what instants are and serve only to bring home the problematic character of instants. Instants and durations alike are abstract items not given in experience. For this reason both durations and instants are philosophically perplexing. But instants are doubly so. For we are apt to be puzzled by the durationless character of instants. While we do have experience of events which last for some time and have some non-zero duration, we do not have direct experience of anything whatsoever that has no duration.

Instants certainly are 'artificial' (Prior) and may be 'superfluous metaphysical entities' (Russell). If we are not to take the courageous line and simply deny the legitimacy of the notion of an instant (and hence to abandon our most cherished theories of the physical world), the possibility of giving some form of reductive analysis is worth exploring. We will consider first some reductive analyses which seek to analyse talk about instants in terms of talk about relatively unproblematic items such as events. Intervals are then introduced as certain sets of instants. In the face of the inadequacies of these attempts I will make use of Tarski's version of Lesniewski's *mereology* in giving a reductive account of instants in terms of intervals.

It is sometimes possible to 'reduce' items of a certain kind by displaying them as equivalence classes of certain other kinds of items. Leaving aside for the moment certain relativisitic complications to be considered in chapter VIII, the simultaneity relation between events is an equivalence relation. Hence, we might try to construe the notion of an *interval* of time as an equivalence class of events under the simultaneity relation, that is, as a set of simultaneous events. This is plausible since we standardly identify periods of time by citing events which occupy that period. And a certain period of time identified by event E_1 is the same period of time as the period identified by E_2 if and only if E_1 and E_2 are simultaneous. But this will not do as it stands for a variety of reasons. And if the argument of chapter II is cogent it will not do at all. Could something analogous serve as a first shot at instants?

127

Events of which we have direct experience are events that last some finite period of time. Equivalence classes of such events at best are intervals and not instants of time. In the quest for a reductionistic account of instants at least two avenues have been followed. Some[8] have sought to introduce instants as the equivalence classes under the simultaneity relation of *point-events*. Others[9] have sought to generate instants from events without using the notion of a duration by means of a more complex construction than that involved in taking the simple equivalence classes. We will consider these approaches in turn.

The first approach is highly unsatisfactory. For if we think that the notion of an instant needs clarification we will think equally that the notion of a point-event needs clarification. The use of point-events in constructing instants is a case of theft over honest toil. Indeed, point-events are in effect introduced as those things which when taken in equivalence classes under the simultaneity relation give us instants. It is not satisfactory to regard point-events as ideal items, sets of which provide models of actual events.[10] Point-events are not, strictly speaking, events at all. An event involves a change, and hence involves at least a pair of times and some difference between what is the case at the respective times. Hence, given that point-events are *events*, instants cannot be equivalence classes to them. To do the trick required, point-events must be durationless *states*. This notion of a durationless state is no real advance on the notion which we seek to explicate. It seems more natural to introduce the notion of a point-event via the notion of an instant. Given some event in some physical system lasting for some interval of time, a point-event could be introduced as a *state* of the system at an instant of time.

5 THE RUSSELLIAN CONSTRUCTION OF INSTANTS

Russell's various attempts to construct instants have the merit of at least beginning with events of non-zero duration. I will consider the particular version of his reduction of instants that can be more tersely stated, as the difficulties I find in his approach arise with regard to each of his attempts.[11]

Russell[12] begins by introducing a relation which is to be called the relation of being *strictly before*, to be represented by '*B*'. An event is strictly before another if and only if it ends before the other begins. The relation of being strictly after is defined as its converse. That is,

b is strictly after a if and only if a is strictly before b. The relation of *partial simultaneity* is introduced as follows. A pair of events are said to be partially simultaneous if and only if there is some time at which they are both occurring. This relation will be represented by 'Sxy'. Russell lays down the following postulates:

1. B is a transitive, irreflexive (and hence asymmetrical) relation.
2. $Bxy \to -Sxy$.
3. $-Sxy \to Bxy \lor Byx$.

We define the *initial contemporaries* of an event E as the set of all events partially simultaneous with E which do not begin later than E (i.e., are not strictly after anything partially simultaneous with E), and add the following postulates:

4. An event strictly after some contemporary of a given event is strictly after some initial contemporary of the given event.
5. If one event is strictly before another, there is an event strictly after the one and partially simultaneous with something strictly before the other.

An instant is then defined as an exhaustive class of simultaneous (mutually overlapping) events. *An event is at an instant* if and only if it is a member of the set of events constituting the instant. *An instant is before another* if and only if some member of the former is strictly before some member of the latter. It can be shown that the instants so defined are simply (linearly) ordered by the before relation and that the set of instants is dense.

Clearly these postulates are not satisfied in either the discrete-quantized change world or the fantasy world described in chapter II. For if event E_1 is the beginning of a total vanishing and event E_2 the ending of a total vanishing, there are no events between E_1 and E_2, and hence the fifth postulate is not satisfied. Russell himself regarded this particular postulate as an empirical assumption about the world, an empirical assumption which he saw no reasons for thinking to be true. The only way in which we can ensure, following a construction such as Russell's, that time is dense is to treat Aristotle's principle as a necessary truth. Given that Aristotle's Principle is not a necessary truth we could still regard Russell's postulates as defining an adequate notion of an instant and simply regard some intervals of time as not being composed of instants. This would mean abandoning the goal of reducing intervals to events and once this is given up there seems to be no incentive to seek a reductive account of instants in terms of events. Once we have learned to live with the failure of reductionism for intervals of time there is no reason to insist that instants, if legitimate, must be constructed out of events.

6 INSTANTS AS PROPOSITIONS

First-order formulations of temporal theories involve quantification over instants. Prior has urged it as a merit of tense logic that we can say what we want to say without having to take on any ontological commitments to instants. Thus, he writes:

> What I might call the third grade of tense-logical involvement consists in treating the instant-variables *a, b, c,* etc. as also representing propositions. We might, for example, equate the instant *a* with a conjunction of all those propositions which would ordinarily be said to be true at that instant, or we might equate it with some proposition which would ordinarily be said to be true at that instant only, and so could serve as an index of it. . . . This sounds a highly artificial procedure, but remember that what lies behind it is the belief that 'instants' are artificial entities anyhow, i.e. that all talk which appears to be about them, and about the 'time-series' which they are supposed to constitute, is just disguised talk about what is and has been and will be the case.[13]

A proposition's being true at an instant, *a*, is construed by Prior as its being implied by the universal proposition, *A*, where a universal proposition is the conjunction of all propositions true at the instant. And temporal relations between instants are construed in terms of temporal relations between their constituent propositions.

It is not disputed that it might be fruitful in a formal context to employ either of the devices given in the first-quoted paragraph as *technical devices for labelling instants*. As either an analysis of our notion of an instant or as providing an ontologically preferable alternative to instants, this treatment of instants as propositions is of dubious merit. First, the ontological advantages of this move are far from clear. For the alleged ontologically objectionable quantification over instants is to be replaced by quantification over propositions. Propositions, like instants, are the sort of abstract item that we would either like to dispense with entirely or to analyse in a satisfactory manner. In the absence of either of these it is not clear that this account is a move forward (ontologically speaking). Prior would argue that we must talk about propositions and quantify over them for reasons not related to time and that, as we are stuck with them, it *is* an ontological step forward to replace talk about instants by talk about propositions. However, it is not clear that banishing quantification over instants in favour of quantification over propositions really avoids ontological commitment to instants. For ontological commitment is not a function just of the syntax of a language: it is a function of a *theory*, i.e., a language together with an intended interpretation. Even if a language contains no terms for items of a certain kind, K, the user of that language is committed to the existence of Ks, if, in giving the truth-conditions for the sentences of that language, reference must be made to K-type items. Explaining the truth-conditions for sentences of the form 'Fp' will involve something like the following: 'Fp' is true at some time t if 'p' is true at a time later than t. If instead of talking about things as true at instants of time we talk of universal propositions implying what we would say was true at the instant, and if we assert the appropriate tense-logical postulates instead of saying that instants are dense, we have not avoided an ontological commitment to instants if we have to make reference to instants in giving the truth-conditions of utterances which make no explicit reference to instants. Ontological commitments are not avoided by avoiding names for the items in question in the object language if reference to them is needed in giving the semantics for the language. This sort of objection would not apply to the standard reduction of the rationals to certain equivalence classes of pairs of natural number. This reduction equates any rational number a/b with the set of all ordered pairs of integers (m,n) which are such that $mb = an$. The standard operations on the rationals are defined in terms of operations on the integers. For instance, addition of rationals is defined as follows: $a/b + c/d$ is the set of all ordered pairs of integers (m,n) such that $n(ab + bc) = m(bd)$.[14] Thus, one can give the truth-conditions of $a/b + c/d = e/f$ in terms of conditions on the integers a, b, c, d, e and f without reference to rational numbers. But one cannot give the truth conditions of '$Fp \rightarrow FFp$' without reference to instants. Thus, Prior's analysis is at best only an apparent and not a real elimination of instants in favour of propositions.

Setting aside the question of ontological merit, there are at least two serious difficulties remaining. Let t be some instant of time and let T be the conjunction of all propositions true at time t. Call this proposition the time's *associated universal proposition*. Prior treats a proposition as something that can change its truth-value from context to context. For Prior, the proposition expressed by the sentence 'The grass is now green' said by me in the spring is the same as the proposition expressed by you when you said that sentence last winter.

If times are construed as propositions, a time t_1 will be distinct from time t_2 if and only if t_1 has an associated proposition T_1 which is not the same as the associated proposition T_2 of t_2. But under Prior's notion of a proposition, it is incoherent to talk of a perfectly cyclical history system in linear time. If we can only count times by counting the different universal propositions that are true we cannot have a multitude of times at which the same universal proposition is true. We have already argued in chapter III that it is coherent to talk of a cyclical world in linear time. Prior's account of propositions, together with his account of instants, entails that this is incoherent. As it is not the account must be rejected.

If, then, we accept the argument of chapter III mentioned above, we cannot treat instants as constructs from propositions in this sense of 'proposition'. Neither can we drop Prior's account of propositions and preserve the analysis. For Prior's universal propositions must not contain names for instants as they are intended as devices for dispensing with instants. If we adopt a view of propositions that gets round the cyclical process-linear time problem by treating the proposition expressed by the sentence 'The grass is now green' as something, which if true, is true at all times, we are treating the proposition expressed as being of the form 'The grass is (tenselessly) green at time t' where 't' is the name of the time demonstratively picked out by uttering 'now'. The associated universal proposition of a particular time on this understanding of 'proposition' will contain some device for referring to that time. In the case of the cyclical history-linear time world, the different times at which the same state recurs will have different associated universal propositions as the propositions contain different names denoting the different times. But then the device of universal propositions cannot be used as a means for avoiding ontological commitment to instants.

There is a further difficulty in Prior's procedure. Any reductionist programme according to which items of, say, type ψ, are really constructions out of items of type ϕ, will not be vindicated unless it is established that the constructed items have the properties which we take ψs to possess. For instance, if we define certain equivalence classes of ordered pairs of natural numbers, we have not vindicated the claim that these are the positive rationals unless we have shown, among other things, that there is an operation definable on these equivalence classes which behaves just like the operation of addition of fractions. We think of instants as dense and durationless. To guarantee that instants treated as propositions have these properties we have to adopt certain tense-logical theses. For instance, we need $Fp \rightarrow FFp$ and $Pp \rightarrow PPp$ in conjunction with the postulates of Lemmon's K_t calculus to guarantee density.[15] While these guarantee density, they are not sufficient for us to be able to assert a plurality of instants during a completely change-less period of time. Assume for the moment that we are concerned only with non-metrical tense logic. In that case, we find that during any changeless period the supposition that there is a plurality of instants leads to the consequence that the same set of tensed propositions is true at each instant and consequently the plurality of instants collapses into a single instant on Prior's account. If we allow ourselves the resources of a metric logic, this is not so. For it will be true at one instant and not true at any other that the end of the freeze will occur in Δt units (allowing Δt to be a real number). However, in this case we are quantifying over intervals of time in a manner that precludes the possibility of treating intervals as propositions. As we shall see, instants can be reduced to intervals. So given a commitment to intervals there is no incentive to construe talk about instants in terms of talk about propositions.

Prior notes this sort of problem in considering whether there is a distinction between the end of time and the end of change.[16] Prior uses the thesis $Fp \rightarrow F-Fp$ (call this A) to characterize ending time and claims that the thesis $Fp \rightarrow F(q \rightarrow -F-q)$ (call this B) characterizes the end of change. In K_t the latter is a consequence of the former but not vice versa. The idea of there being a future instant at which every future tense proposition is false is captured by A. B is intended to capture the idea that there is a future instant of time, thought of as the end of change, at which some propositions are true but no proposition that is then true will become false. At such an instant some future tense propositions are true but nothing that is true will change its truth value. Things may be true in the future and if they are they will be permanently so. However, if, as Prior notes, we add the assumption that no instant is in its own future,[17] we can show that B is a consequence of

A and hence that *A* and *B* amount to the same thing. Given my argument of chapters II and IV, I have to allow that there is a distinction between ending time and the end of change. I granted that some sense could be made of this distinction but I argued that having posited the end of change the posit of further empty time would be purely gratuitous. However, even if such a distinction can be made, it cannot be made in the way that Prior attempts. For even if we can make sense of the notion of an instant's being in its own future in the context of *closed time*, no sense whatever can be made of this supposition in the context of *non-closed time*, which is the supposition that we are making in entertaining either the idea of an end to time or an end of change. Prior is forced to attempt to attach a sense to ending change in non-ending time through the desperate move of talking about instants in non-closed time as being in their own future because his technique for individuating instants is not sufficiently strong. Prior claims that to say no instant is in its own future is to beg the question of the possibility of an end to change in the non-ending time. But that is to misconstrue the nature of the objection. Once we see that accepting his distinction requires us to countenance instants in their own future in non-closed time, our doubts about the sense of that transfer to the sense of his distinction. It is no help to add, as he does, that there are consistent tense logics of circular time in which every instant is in its own future.

7 INSTANTS AS PARTS OF DURATIONS

In what follows I will assume the notion of a duration or interval of time and will use it to explicate the notion of an instant with the help of Tarski's version of Lesniewski's mereology.[18] While one can explicate the notion of an instant in terms of that of an interval, one can equally explicate the notion of an interval of time by reference to the notion of an instant. An interval might be construed as a composite object formed from instants. In some systems it is to some extent arbitrary which notions you treat as primitive and which as derived. For instance, in geometry we may prefer to treat lines as sets of points; or points as the intersection of lines. Which choice we actually make should be governed by reference to what is, for the purposes at hand, the most enlightening thing to do. The reason for preferring to take intervals as more basic is that we are not as puzzled by intervals of time as we are by instants of time. This is so, as noted earlier, if for no other reason, because we have direct experience of events that last for some interval of time and we have no direct experience of anything obtaining for only an instant. Unlike the notion of an interval, the notion of an instant is not a notion which we apply directly to our experience of the world.

The informal construction to be offered of instants in terms of intervals is neutral with regard to *AP* (Aristotle's Principle) in the following sense. Both those who take *AP* to be necessarily true and those who take *AP* to be, at best, contingently true can accept the definition of instants in terms of intervals. The difference between these positions will emerge when the question of the analysis of the notion of an interval arises.

Mereology, as understood by Lesniewski and Tarski, is the general theory of the relation of parts to wholes. Tarski, in the paper cited above, formulated a theory of mereology, the postulates of which provide those truths about the relations between parts and wholes of things which obtain regardless of the kind of thing in relation to which we are talking of wholes and parts. The basic postulates are intended to be true if we interpret the whole–part relation in terms of spatial inclusion, temporal inclusion and so on. Tarski adds certain further postulates which relate specifically to spatial parts and wholes in order to develop three-dimensional Euclidean geometry of solids. By taking the whole–part relation and the notion of a solid as basic he is able to define the notion of a point in terms of the notion of a solid.

An interval is an extent of time and the central idea behind the construction to be offered is the following. Thinking of something as an interval of time commits one to thinking of it as having temporally extended parts. I shall first give a formulation of some aspects of this notion of being 'a temporal part of'. Then, using some notions of temporal order, I shall offer a definition of an instant. The point of my procedure is to show that if we are happy to talk about intervals of time, instants can be had as an ontologically harmless consequence of our manner of talking about intervals. While we will not want to define intervals as equivalence classes of events, no harm will come in following the argument if one thinks for intuitive purposes of an interval in this way, and thinks of particular events as representatives of the equivalence classes. In referring to some postulates as mereological postulates it is implied that these should hold when the relation of 'being a part of' is interpreted in domains other than that of intervals of time.

In what follows, lower-case letters are used as names or variables for intervals and parts of intervals. Upper-case letters are used as variables or names for sets of intervals. We introduce the two-place predicate letter '*Pxy*' which is intended to stand for the notion of being a part of. Our first mereological postulate is the following:

MP1:   *P* is a transitive, asymmetrical relation.[19]

The second mereological postulate formulates the basic idea that anything extended has parts that are themselves extended:

MP2: $(x)\,(Ey)\,(Pyx)$.

We introduce the two-place predicate '$x < y$' which is intended to represent the notion of being entirely before, in the sense of starting before and ending before with no temporal overlap. The properties of this relation are given in the first temporal postulate:

TP1: $<$ is a transitive asymmetrical relation.

The statement of our other temporal postulates will be expedited with the aid of the following definition. We will say that x is *entirely after y* if and only if y is entirely before x. We will say that x is *partly before y* if and only if some part of x is wholly before y. Similarly, x is *partly after y* if and only if some part of x is entirely after y. We will say that an interval x is *adjacent to an interval y* just in case x is entirely before y and there is no interval entirely after x and entirely before y, and there is no interval entirely before y having a part partly after x. We write '*Axy*' for 'x is adjacent to y'. The second and third temporal postulates are designed to rule out temporal 'gaps'.

TP2: $(x)\,(Ey)\,(Axy)$.
TP3: $(x)\,(Ey)\,(Ayx)$.

We introduce definitionally the notions of initial part, I, and final part, F, of an interval. An *initial part* of an interval is a part of an interval such that no other part of the interval is entirely before it. A *final part* of an interval is a part such that no part of the interval is entirely after it. The fourth and fifth temporal postulates specify that an interval has an initial part and final part.

TP4: $(x)(Ey)(Ixy)$.
TP5: $(x)(Ey)(Fxy)$.

By an inner part of an interval, C, we mean a part of an interval such that there is a part of the interval entirely before that part and a part entirely after that part such that the sum of the three parts equals the interval. The next temporal postulate asserts the existence of an inner part:

TP6: $(x)(Ey)(Cyx)$.

We now proceed informally to define a notion of an instant within this framework. By TP4 any interval has an inner part and that inner part in virtue of being an interval has itself an inner part and so on. Hence we can introduce instants as *infinite nested sequences of inner parts of intervals*. This means in effect that an instant is an infinite set of Chinese boxes (the intervals). Each box has a finite non-zero size but for any box there is inside it a still smaller box. That is, using lower-case letters i, j, k as variables for instants, i is an instant if and only if i is a set of intervals such that (1) for any pair of members of i, x and y, either x is an inner part of y or y is an inner part of x *and* (2) for any x in i there is an inner part of x in i. Consequently any interval has associated with it a dense set of instants.

We require one further temporal postulate which it is convenient to formulate with the aid of our definition of an instant. This is the following:

TP7:   If i is an instant then there is no interval x such that x begins after and ends before every member of i.

The next stage in a full development of this sort of account of instants would involve showing that instants so defined have the properties we wish them to have. As the direction of such a development is, I think, fairly clear, I shall not here do anything more than sketch a proof that instants so defined are dense. To this end we define an ordering relation, B, such that Bij if and only if some member of i is strictly before some member of j. We can see quite simply that B is a dense ordering. Let i, j be distinct instants and suppose Bij. Then for some x_i in i and x_j in j, $x_i < x_j$. Now x_i and x_j have inner parts $x_i{}', x_j{}'$. Hence, there is some part x'' of x_i such that $x_i{}' < x''$ and $x'' < x_i$. Consider x''. x'' generates an instant $i_{x''}$ — the nested sequence of intervals such tnat a member x'' of that sequence is strictly before a member x_j of j. And a member x_i' of i is strictly before a member x'' of $i_{x'}$. Therefore, $i_{x''}$ is an instant such that $Bii_{x''}$ and $Bi_{x''}j$. Additional constraints on the system of intervals are required if the system of instants defined by the procedure is to be continuous.[20]

8 INTERVALS OF TIME

The conclusions that have been reached with regard to intervals of time have been limited and largely negative. It was suggested that the notion of an interval should be taken as more basic than that of an instant, and it has been argued that neither a reductive account of intervals in terms of events nor one in terms of a proposition is viable. There are no other items in terms of which it would be *prima facie* plausible to construct intervals. Consequently we should recognize that intervals of time are a species of abstract object and a philosophical adequate theory of intervals will involve providing an account of how we identify intervals of time and providing a criterion of identity for intervals. In developing an account of intervals of time as abstract objects it will be enlightening to have available a brief characterization of our paradigm of an abstract object. This in turn can best be developed through a contrast with our paradigm of a concrete object.

Our paradigm of a concrete object is a medium-sized macroscopic item located in space and time which is a possible object of human perception and which is consequently capable of being non-verbally ostended. Such an object is a possible subject of change and as such is capable of entering into causal interaction with other objects. By contrast a paradigm abstract object has the following characteristics:

1 It is a necessary but not sufficient condition of an object's being an abstract object that it cannot be a possible object of human perception.
2 Consequently an abstract object cannot be non-linguistically ostended.
3 Thus, the identification of abstract objects must be done by linguistic means. Standardly, this is achieved by introducing a functional expression such as 'the weight of x'. In the case of abstract objects there is no way of identifying the object except as the reference of such a phrase.

4 Abstract objects cannot enter into causal relations. And, in general, there are no contingent truths about abstract objects that are not best construed as contingent truths about the objects referred to by the referring expressions contained in the functional expressions which identify the abstract objects in question.

Abstract objects as diverse as weights and numbers satisfy the above characterization. In what follows it will be fruitful to follow Dummett[21] in delimiting a sub-class of abstract objects, to be called *pure abstract objects*. This distinction is drawn by noting that terms for *some* abstract objects are related in their use to objects given in perception. For instance, we can only make a successful reference to a weight by making reference to some actual non-abstract object or objects. On the other hand, if one has a Fregean conception of number, one will think that while reference may in fact be made to numbers by citing concrete objects, reference can also be made to numbers in a way that does not presuppose reference to any concrete object or objects. Such abstract objects whose existence is independent of what concrete objects there are will be called *pure abstract objects*.

Temporal items such as intervals are clearly not possible objects of human perception (characteristic 1). Our means of identifying temporal items is linguistically dependent. For we standardly identify such items either by citing some event, happening, process, etc., or by using temporal indexicals to give a demonstrative identification of a time, as when, for example, I refer to this moment *now*. However, in either case the temporal items satisfy the condition of being identifiable only by non-ostensive linguistic dependent means (characteristics 2 and 3). Furthermore, we do not normally think of temporal items as entering into causal relations. And *many*, but not all, contingent truths apparently about temporal items (i.e., the time of the battle was before the time of the festivities) can be construed as a contingent truth about the events cited in identifying the temporal items in question (i.e., the battle occurred before the festivities) (characteristic 4). If this was really so, one might take the line that we have no need to suppose the existence of real referents corresponding to terms for temporal items in true assertions, on the grounds that the truth-conditions of any assertion about temporal items can be given without requiring us to suppose that there are such objects. That is, if we can explain the meaning of all assertions in which reference is made to temporal items in terms of assertions in which no such reference is made, we have no reason to

139

suppose there really are such temporal items. Of course, one can say that there are such items but this will be construed in such a manner as to display these items as merely a by-product of our manner of speaking. This manner of construing the referent of terms for temporal items will be called the *reductionist construal*. However, the possibility of time without change shows this conception of temporal items to be untenable. We can envisage contexts in which we would regard ourselves as warranted in making assertions such as: 'Between the time I picked up my glass and set it down, one year of time devoid of all change passed'. We cannot give the truth-conditions of such assertions via a biconditional which on one side makes no reference to temporal items.

It is enlightening to consider a mathematical analogy. If the only contexts in which we employed numerals were contexts in which we gave the number of things falling under some empirical predicate such as '. . . swan(s) in the pond', one could maintain that as the truth-conditions for such assertions can be given in terms of first-order quantifiers and identity, one has no need to suppose that there really are objects answering to the numerals. For example, the meaning of the sentence 'The number of swans in the pond is 2' would on this account be rendered more perspicuous by the sentence 'There is a swan x in the pond and there is a swan y in the pond, x is not the same swan as y and nothing is a swan in the pond unless it is x or y'. Similarly, if the only contexts in which we talked about temporal items were contexts in which we identified the temporal items by means of events occurring at or during the temporal time (i.e., 'An hour passed while Icabod slept'), we could resist positing temporal items on the ground that that claim could be explained in terms of assertions making reference only to events (i.e. the sun moved through such and such a distance while Icabod slept). However, in the case of numbers there appear to be sentences such as '2 is even' which cannot be handled in the above manner. The truth of such sentences has inclined some to view numerical words as having real referents which cannot be construed in a reductionist manner. Analogously, the possibility of contexts in which reference to temporal items cannot be handled as above may incline us to posit the real existence of temporal items. I will call any view that incorporates the following two claims a *realistic construal* of the reference of terms for temporal items. First, there are temporal items and reference must be made to such items in explaining the truth-conditions of some assertions about time and the temporal aspects of things. Second, assertions about such items cannot be construed reductionistically.

The view of time that has been called Platonism involves a realistic construal of the referents of terms for temporal items. In holding that the time system existed independently of the history system the Platonist is construing temporal items as *pure* abstract objects. The Platonist allows that one in fact identifies a temporal item by citing some event occurring at that item. However, he holds that the existence of temporal items does not presuppose the occurrence of events which *we* would need to cite in identifying particular temporal items. It is possible to be a realist in the sense outlined above without being a Platonist. One might conclude on the basis of the possibility of time without change that a realist rather than a reductionist construal must be given to the referents of terms for temporal items. However, one could hold that temporal items are abstract objects but not pure abstract objects. In this case one would hold that there could be no time system without the existence of a history system. Given that there is a history system, the realist will regard it as reasonable for the inhabitants of some possible worlds to believe in the existence of a richer system of temporal items than the reductionist can allow.

It is of interest that the contexts which require a non-reductionist treatment of temporal items are contexts in which temporal items are vested with limited causal powers. For as we saw in chapter II, any context which would prompt reference to temporal items during which no events occurred would be a context in which causal powers would have to be ascribed to the passage of that period of time. In this case temporal items fail to satisfy what was given as the fourth characteristic of a paradigm abstract entity. This suggests that we should view temporal items as being in some sense *sui generis*. While temporal items cannot be seen as concrete objects they also cannot be seen to fit perfectly our paradigm of an abstract object.

It is not implausible to offer a similar view to spatial items. One might reject a strictly reductionist theory of space on the grounds that one wanted to admit the possibility of spatial vacua. At the same time one might wish to treat spatial items as abstract items but not as pure abstract items. As with temporal items, one can envisage contexts in which one would wish to ascribe causal powers to spatial items. Thus one would be led to see spatial items as failing to satisfy the fourth characteristic of our paradigm of an abstract object. There are those who think that we should treat space and time together and offer a theory which has as its basic notion the notion of a *spacetime*. One could reason along lines similar to those outlined to reach the conclusion that the referents of terms for spacetime should be treated realistically. Such a conception of spacetime items seems particularly appropriate to the project of geometrodynamics, a consideration of which is not possible within the confines of this book. The project of geometrodynamics is to give a unified physical theory which is a complete and purely geometrical description of the physical world in the sense that the contents of the world would be represented by the varying curvature of space, and change would be represented by variations in time of that curvature. The presence of a macroscopic physical object would be construed as a certain curvature in the manifold of spacetime points. Thus, spacetime items would have a real existence, being the ontologically most basic items of the physical world. While there may be difficulties in this project, it is plausible enough that it should count in favour of the theory of time that it is compatible with this theory.[22]

Intervals of time, I have suggested, should be treated as abstract entities sharing most but not all properties of our paradigm of such an entity. Their identification is parasitic on the identification of items in time and there is no reason to suppose that such intervals of time would exist in the absence of things in time relative to which they can be identified. Thus, while there may be intervals of time not containing events, there is no reason to assume that a world entirely empty of all change would none the less contain intervals of time.

In addition to providing an account of the identification of a type of abstract time, a philosophical adequate account of such items should include a specification of their conditions of identity, which in the case of intervals of time can be given as follows: an interval of time i is the same interval as an interval j if and only if *either* the identifying events for i and for j are simultaneous or in the case where i, j are periods of empty time i and j bear the same *temporal relations* to the events which are used in identifying i and j. For instance, if i is identified as the year between the disappearance of x' and the reappearance of x', and j is identified as the year between the disappearance of x'' and the reappearance of x'', then i is the same interval as j if and only if the disappearances of x' and x'' and the reappearances of x' and x'' are respectively simultaneous. The force of the qualification 'the same temporal relations' is to allow relations of identity and diversity between parts of some period of empty time. That is, the first month of a freeze is identified, in a sense, by reference to the same events as the last month of the freeze, but are individuated by means of the differing temporal relations they bear to these events.

VII

THE METRIC OF TIME

Time has no divisions to mark its passage, there is never a thunder-storm or glare of trumpets to announce the beginning of a new month or year. Even when a new century begins it is only we mortals who ring bells and fire off pistols.
Mann, 1960

1 DATES AND DURATIONS

In the preceding six chapters we have been occupied by questions concerning temporal order. There is of course much more to time and the temporal aspects of things than has been captured by our account so far. To see this we need only reflect how little we would know of time and its contents if we knew only the topological structure of time and the order in time of its contents. In addition we wish to have the answers to questions of the form: how long ago did such and such an event occur? how long did it last? One answer to the latter sort of question which would be appropriate in some contexts would be that the event lasted as long as some other event. However, in order to do physics we need to establish a framework within which we can answer such questions by assigning a numerical value to the duration of an event. A framework which allows us to give this sort of answer to this and related questions will be called a *metrication of time*. My aim in this chapter is to develop a philosophically adequate understanding of what such a framework involves. Measuring time involves the application of mathematical items, i.e., numbers, to non-mathematical items, i.e., intervals of time and events. Consequently, a study of the measurement of time requires a study of the aspects of the system of non-mathematical items which are to be given a mathematical representation, a characterization of the mathematical system to be used in the representation, and an account of the relation between the non-mathematical and the mathematical systems. In this section we will be primarily concerned with the first of these; the remaining two will be discussed in the next section.

Basically what we wish to achieve through a metrication of time is a systematic way both of assigning dates to things in time and to temporal items and of assigning measures to their respective durations. Perhaps the simplest notion involved is that of a *date*. But just what is a date? In saying, for example, that the date of the next college meetings is 17 January we are identifying a period of time in a systematic fashion. Thus, we can think of a date as a temporal item identified so as to allow us to determine from the mode of identification of that item its temporal relations to another item dated in the same system. The relatively simple rules of our conventional dating system allow us to infer, for example, that the time labelled '1 January 1974' comes before the time labelled '2 February 1975'. Thus, a dating system provides a systematic way of labelling times where the labels are attached in a rule-governed way that reflects the order of the labelled times.

If we were only interested in giving the order of a sequence of intervals or periods of time that come before one another without temporal overlap, it would be sufficient to attach a single label to the periods by associating a sequence of integers, say, to the sequence of intervals. However, there are other relations that can obtain between intervals of time which we wish to represent in our dating system and whose representation will require assigning a pair of labels to the intervals. One of these relations is what will be called the relation of *partial beforeness* which obtains between a pair of intervals I and J just in case I begins before J begins and ends before J ends and part of I is simultaneous with part of J. The other relations are those of *inclusion* and *proper inclusion*. An interval I includes an interval J just in case J does not begin earlier than I and J does not end after I. The relation of proper inclusion holds just in case I begins after J and ends before J. In the case of these relations a pair of labels is needed to date the beginnings and ends of the intervals.

As we are interested in the mathematical representation of the metrical aspects of time, we will understand a dating system to involve the

144

assignment of mathematical items to intervals and events. A dating system that represents the relations of being before, being partially before, being included and being properly included will be said to be *topologically adequate*. The notion of representing can be explicated as follows. Let I and J be intervals of time which may have been identified as the intervals during which the events E_1 and E_2, respectively, have occurred. A dating system represents the relation of, say, inclusion, if there is some mathematical relation which holds between the mathematical items assigned to I and J whenever I includes J. This means that if we have a topologically adequate dating system we can tell from a correct assignment of mathematical labels to a pair of intervals which of these relations hold between the intervals in question.

The relations we have been considering have not been taken to be particularly problematic from the epistemological point of view. For if we set aside relativistic considerations (which will be taken into account in the next chapter) and restrict our attention to intervals identified by events occurring in the same vicinity, we can see by observation in many cases which of these relations hold between the events in question and hence which of these relations hold between the intervals identified by reference to these events. We see, for instance, that my striking of a particular match began after I started to lecture and ended before I finished lecturing. Hence, the interval identified by the striking is seen to be included in the interval identified by the lecturing. In contrast to these relations there are three relations of utmost importance to the metrication of time which have traditionally been seen as epistemologically problematic. These are the relations of lasting as long as, lasting longer than, and lasting p/q as long as (where p and q are positive integers). For with the exception of cases where one interval includes another and, hence, is obviously longer with respect to duration, philosophical controversy wages as to what, if anything, grounds the judgments that, say, one event lasts as long as another. Here and in what follows, I will assume for ease of exposition that we are concerned only with intervals of time identified by reference to events occurring during the intervals, and will formulate the account of metrication being developed in terms of events and not intervals. Modifying the account to cover temporal vacua will require only minor reformulations.

How do we determine whether or not two events, neither of which encloses the other, have the same duration? At first glance this may seem particularly problematic because we implicitly compare this question with the analogue for space. How do we discover whether one object is longer than another? We may seek to determine this by placing the two objects in juxtaposition with one end of each aligned. We can then simply see whether or not the other ends are aligned. In the case of time we cannot take two successive events and lay them side by side so that they start together with a view to seeing whether they finish together. This can make it look as though determining the extension of the equivalence relation of lasting as long as (that is, determining which pairs of events last as long as each other) is epistemologically a more problematic matter than determining the extension of the equivalence relations of being as long as. Such an impression is quite misleading. For the unproblematic spatial case where we have two objects together in space has its analogue in the unproblematic temporal case of a pair of events together in time one of which includes the other. To compare two spatially separated objects we can either bring the objects together or use some third object which is compared to each of them. In either case there is motion of an object through space. If objects were to change length under transportation through space this procedure for determining the relative lengths of objects when separated would not be acceptable. Thus, in using the procedure of transporting a rigid rule we are assuming the length of the object to be invariant under spatial transportation. The question arises as to how to ground that assumption. In the case of time, the standard technique for determining the relative duration of successive events is to use clocks. However, this procedure would not give us the relative duration of temporally successive events if the rate of the clock changed with time. Thus, in using this procedure we are assuming that our clocks generate a sequence of events having the same duration. And the question arises as to how, if at all, we can ground that assumption. Later in this chapter we will take up this vexed issue. However, for the moment it will be assumed that we have identified some physical system, our clock, which produces a sequence of events which we take to be equal with regard to duration, or, as we shall say, are *isochronic*. This assumption whose justification is problematic has a spatial counterpart (the assumption that rigid rods preserve their length under spatial transportation) whose justification is also epistemologically problematic. It is, contrary perhaps to one's first impression, not the case that the measurement of time is more problematic than the measurement of space in this regard.

As we have noted, the straightforward answer as to how we compare events as to relative duration is to use a clock. If my clock is a pendulum and if my last lecture lasted between n and $n + 1$ swings of the pendulum, and my colleague's lecture lasted between m and $m + 1$ swings of the pendulum, my lecture was not as long as his if n is less than m. If n is greater than m, then mine is the longer lecture. If m equals n, the lectures cannot be discriminated by reference to the pendulum with regard to duration. While the lectures may not be discriminably different with regard to duration using this particular technique, it may well be that a finer clock, a clock generating shorter events, will allow us to discriminate between the events with regard to duration. This point illustrates the fact that in using these operational techniques, we cannot determine that one event lasts exactly as long as another. We ascertain only that one event is not distinguishable in regard to relative duration from another. While the relation of lasting as long as is an equivalence relation, the relation of not being discriminably different in regard to length is not. It is reflexive and symmetrical but it fails to be transitive. It may be that we can discriminate with regard to duration neither between events E_1 and E_2 which begin together, nor between events E_2 and E_3 which begin together, but that we can distinguish between E_1 and E_3. Consequently, we take the judgments we make on the basis of these operational techniques as being *defeasible* in the following sense. If we cannot discriminate between a pair of events in regard to duration we assume that they are equal in duration. However, we would revise this assumption if it were to lead to conclusions which violate the transitivity of the relation of being equal with regard to duration.

We wish to use our clock not only to determine whether one event lasts as long as another, but also to assign a measure to the duration of any given event. If all events lasted as long as some integral number of clock events we could simply take as the measure of the duration of an event the number of events in a sequence of clock events (events produced by the clock) which lasts as long as the given event. However, this is not in general the case and, consequently, the notion of one event lasting p/q as long as another event (where p and q are positive integers) is crucial to the establishment of a metrication for time.

In explicating the notion of one event's lasting p/q as long as another it will be fruitful to begin by considering the notion of one object's being p/q as long as another. The motivation to use fractions in giving the length of objects arises from having to deal with objects the length of which is not an integral multiple of the length of our rigid rule. Given, say, an object shorter than the rule we can find positive integers p and q such that laying down the rule q times marks out a spatial interval of the same length as that marked out by laying down the object p times. p/q gives the ratio of the length of the object to the length of the rule. So to say that the length of an object is p/q is to say that laying down the object q times gives a length equal to that obtained by laying down the rule p times. Similarly, we can explain what is meant by saying that one event E_1 lasts p/q as long as an event E_2 as follows: a sequence of q isochronic occurrences of events of the type of E_1 lasts as long as a sequence of p isochronic occurrences of an event of type E_2. Thus, we can think of an assignment of a rational value to the length of an event as giving the ratio between the duration of that event and the duration of an event produced by our clock.

While the technique outlined constitutes an explanation of what we mean when we assert that the length of an event is p/q, it is not in general the technique we actually employ (as will be indicated below). However, we could use this technique in certain circumstances and for the sake of illustration let us suppose we are applying it to an event that has less duration than a clock event. In this case we find the numbers p and q such that p occurrences of the type of event in question have the same duration as q occurrences of the clock event. But if we are using our clock to determine which events last as long as each other under the assumption that the clock events are isochronic, how do we know that the p occurrences of the event being timed are isochronic? If they are not, this is not a viable technique for determining the ratio in question. A similar problem arises in the case of spatial measure. The technique outlined is only a viable way of determining the ratio of a given object's length to the length of the rule if the length of both the rule and the object being measured are not affected by spatial transportation. In the spatial case we simply assume that whatever it is, if anything, that licenses the assumption that the rule is unaffected in regard to length by transportation also licenses the assumption that objects measured are similarly unaffected. And in the absence of information concerning disturbing influences which would affect the isochronic character of our clock, we assume that the sequence of events of the type of the event being timed are isochronic. Thus, we see that in the case of spatial measurement we have to assume that objects are unaffected by transportation and not just that our rule is unaffected. Similarly, in the case of temporal measurement we have to assume not only that the clock generates isochronic events but that the events to be timed can be isochronically reproduced. This point, which may seem obvious, needs to be remarked upon as many write as if measurement only requires the assumption that clock events are isochronic. However, as we have seen, in using the technique under consideration we have to make a general assumption about the isochronic character both of the timer and the timed.

The technique that has been outlined is not the one which we generally use in applying fractions to the duration of events. We are more likely, for instance, to use as a clock a physical system in which a pointer rotates around a fixed point. The clock events in this case are complete rotations and by geometrically graduating the circle into parts of equal spatial length, we assign to an event the measure p/q where p/q gives the ratio of the length marked out while the event occurs to the length marked out by the clock event. We do not require repeated occurrences of the event being measured and hence we avoid the problem of justifying the assumption that the repetitions are isochronic. However, we have now to consider the assumption that each event of the pointer's sweeping out an equal spatial path has the same duration. And, obviously, this assumption is not a consequence of the assumption that each complete rotation is isochronic. So the general point remains. More is presupposed in measurement than merely the isochronic character of clock events. While this last technique is most commonly encountered in determining the ratio of the length of an event to the length of a clock event, the other technique is more fundamental to an understanding of what is meant by saying that one event lasts p/q as long as another. For we can and do employ this notion in regard to, say, digital clocks where we cannot use spatial graduation in determining the ratio of an event's duration to the duration of a clock event.

It should be noted that in determining the ratio of the duration of one event to another we will never have occasion to use anything except the positive rational numbers. For, given any events E_1 and E_2, we will be able to find positive integers such that a sequence of p occurrences of isochronic events of the type of E_1 are not discriminably different with regard to duration from a sequence of q occurrences of isochronic events of the type of E_2. Consequently, one who assumes that for any event E there is some positive rational giving the ratio of that event to a clock event cannot be refuted by the outcome of any attempt to ascertain the ratio directly. Certainly one who opted for a particular value, p/q, for a given event might have to revise his assumption given finer techniques for determining whether events begin and end together. However, while this might lead him to revise the particular rational value assigned it would not show that there is no rational value of the ratio. In view of this, I will assume that for any pair of events there is some rational number giving the ratio of their durations. One who makes the stronger and empirically unjustified assumption that there may be events the ratio of whose duration is not a rational ratio can regard the operational assignment of rational ratios through direct measurement as being, in some cases at least, only an approximation to the true ratio which cannot be represented by a rational value and which consequently cannot be ascertained through direct measurement.

We noted that an adequate metrication of time must provide not only a topologically adequate dating system, it must also provide a systematic assignment of measures of duration to all events and intervals of time. Any system which assigns measures of duration so as to represent the relations of lasting as long as, lasting longer than and lasting p/q as long as will be said to be *metrically adequate*. Letting E_1 and E_2 be arbitrary events and letting $m(E)$ represent the measure of duration assigned to event E we can express this condition on m as follows:

m is metrically adequate if and only if
1. $m(E_1) = m(E_2)$ if and only if E_1 lasts as long as E_2;
2. $m(E_1) > m(E_2)$ if and only if E_1 lasts longer than E_2;
3. $m(E_1) = p/q \, m(E_2)$ if and only if E_1 lasts p/q as long as E_2.

An adequate metrication will involve a topologically adequate dating system which assigns to event E some mathematical item $d(E)$. For instance, the dating system might assign to E an interval of rational numbers from and including a to and including b which we write as $[a,b]$. The metrically adequate measure system employed in the metrication will assign some value $m(E)$ to the event E. In addition, in an adequate metrication the dating system and the measuring system must inter-relate in the following way. There will be some function, f, defined on the items assigned to events by the dating system whose value for the item, $d(E)$, assigned to E will equal the value, $m(E)$, assigned to E by the measure system. For instance, if the dating system assigns intervals of rational numbers to events, the function f might assign to $[a,b]$ the absolute value of the difference between an a and b, $|b - a|$. This is $(b - a)$ if b is greater than a and is $-(b - a)$ otherwise. To meet the condition on the inter-relation of the dating system and the measure system these systems have to be set up so that $m(E) = |b - a|$. This intuitive characterization of what is involved in an adequate metrication

150

of time will be supplemented in the next section by the development of a more rigorous account in the course of which it will be shown how one can arrive at a metrication which satisfies this constraint on the inter-relation of the dating and measure systems.

2 THE METRICATION OF TIME

A metrication of time involves a mathematical representation of certain aspects of time and the system of all events. One aspect of time to be represented is its topological structure. Thus, in selecting a mathematical system, items of which are to be assigned to temporal items and things in time, we need to select one which has the same topological structure as time. For the sake of illustration it will be assumed that time has the standard topology; that is, time is non-beginning, non-ending, continuous and linear. Under this assumption an appropriate mathematical system is the real number system as this system has the properties in question. Our topologically adequate date system must assign mathematical items to events so as to represent the relations of being before, being partially before, being properly included by and being included by. This constraint on the dating system can be met if we define a function d which assigns to an event an interval of real numbers. The intervals to be assigned will be one of four types: open–open, closed–closed, open–closed, closed–open. We write (a,b) for the open–open interval consisting of all real numbers between a and b. $[a,b]$ is the closed–closed interval consisting of all a and b and all real numbers between a and b. The open–closed interval $(a,b]$ consists of the real number b and all real numbers between a and b. $[a,b)$ the closed–open interval, consists of the interval constituted by a and the real numbers between a and b.

A dating system with function d that assigns intervals of real numbers to events will be topologically adequate if it meets the following conditions:

1. E_1 is before E_2 if and only if every real in $d(E_1)$ is less than every real in $d(E_2)$.
2. E_1 is partially before E_2 if and only if $d(E_1)$ and $d(E_2)$ have a real in common and some real in $d(E_1)$ is less than any real in $d(E_2)$.
3. E_1 is included in E_2 if and only if $d(E_1)$ is a sub-interval of $d(E_2)$, ($d(E_1)$ may be $d(E_2)$).

151

4. E is properly included in E if and only if $d(E_1)$ is a proper sub-interval of $d(E_2)$.

The inter-relation constraint requires that there be some function, f, defined on the items assigned by d to events (in this case intervals of real numbers) whose value for the item $d(E)$ assigned to E by d must equal the measure $m(E)$ assigned by the measure system to E. As we will see this condition can be met if we let the function f assign to an interval (a,b), $[a,b]$, $[a,b)$, $(a,b]$ the value $|b - a|$. In the case of intervals of any one of these four types we will say that a and b are the *endpoints* of the interval. The function f can be described as assigning to any interval the absolute value of the difference of its end-points. The inter-relation constraint will be met if the absolute value of the end-points of the interval $d(E)$ assigned to E equals the value $m(E)$ assigned to E by the measure function; that is, given $d(E)$ has end-points a and b, $m(E) = |b - a|$.

In this abstract characterization of a metrication of time we will make the highly idealizing assumption that we have a clock which has always been, is and will always be producing events, the sequence of which we label $\ldots c_{-2}, c_{-1}, c_0, c_1, c_2, \ldots$. Further, we take it that these events do not overlap and that there are no gaps between them. Topological adequacy requires that the function d of the dating system assign either a sequence of closed–open intervals or a sequence of open–closed intervals to the clock events. This choice is arbitrary and we will use closed–open intervals. We cannot use closed-closed intervals for that represents the clock events as overlapping for an instant. Nor can we use open–open intervals for that would represent the clock events as having a gap of an instant between them. As we deem the clock events to be isochronic, the measure system assigns the same value c to each event. This means that we must assign to any given clock event an interval $[a,b)$ such that $|b - a| = c$. Consequently, we can meet the topological, metrical and inter-relation constraints for clock events by selecting some clock event, our origin event, c_0, and assigning it an interval $[a,b)$ where $|b - a|$ equals c. Succeeding clock events $c_1, c_2, \ldots$ are assigned the sequence of intervals $[b,2b - a)$, $[2b - a,3b - 2a) \ldots$. Preceding clock events $\ldots c_{-2}, c_{-1}$ are assigned the intervals $\ldots$ $[3a - 2b,2a - b)$, $[2a - b,a)$. For an arbitrary clock event c_n, $d(c_n) = [nb - (n - 1)a,(n + 1)b - na)$. If c_i is assigned $[a',b)$ and c_j is assigned $[a,b')$ the sequence of successive clock events $c_i, \ldots c_j$ will be assigned the interval $[a',b')$.

To have an adequate metrication we must extend the dating function d to assign intervals to events other than clock events. To this end we introduce the notion of the *cover* of an arbitrary event E as a sequence of clock events $c_i, \ldots c_j$ such that c_i begins with or before E and ends after E has begun and such that c_j begins before E ends and ends when or after E ends. d will assign some interval $[x,y]$ to this sequence of clock events. We extend d to E by assigning a sub-interval (not necessarily a proper sub-interval) with end-points $[x',y')$ of that interval to E subject to the following constraints. First, the totality of such assignments to all events is topologically adequate. Second, the absolute value of the difference of the end-points, $|y' - x'|$, equals the value p/q where p/q is the ratio of the duration of the event to the duration of a clock event determined by the measure system. Any such extension of d to the set of all events generates an adequate metrication of time.

In setting up a metrication there are at least two sources of arbitrariness. First, given that we have a metrication M which assigns intervals of length c to clock events and assigns c_0 the interval $[a,b)$, we can generate an adequate metrication M' which assigns to c_0 some other closed–open interval $[a',b')$ of that length. To effect the translation from M to M' we simply add to the end-points of the intervals assigned by M the constant $a' - a$. Second, given metrication M we can generate another adequate metrication M'' which assigns intervals of a different length to the clock events. If, for example, M assigns intervals of length of say 1, and M' assigns intervals of length k to clock events, we can translate from M to M' by multiplying the end-points of the intervals assigned by M by the factor k. In general we multiply by the ratio, $\dfrac{|b' - a'|}{|b - a|}$, of the measures of the duration of the clock events. Combining these two degrees of arbitrariness we obtain a family of adequate metrications which relate to one another as follows: if M and M' are metrications in this family and M assigns to c_0 the interval $[a,b)$ and M' assigns to c_0 the interval $[a',b')$, and E is assigned $[t_1,t_2)$ by M and $[t'_1,t'_2)$ by M' then:

$$[t'_1,t'_2) = \left[\frac{|b' - a'|}{|b - a|} \, t_1 + (a' - a), \; \frac{|b' - a'|}{|b - a|} \, t_2 + (a' - a) \right)$$

writing a for the constant a'/a and b for the constant $\dfrac{|b' - a'|}{|b - a|}$ we can express this in terms of the end-points by writing: $t' = at + b$. A transformation rule of this form is a *linear transformation* and we can express the conclusion reached by saying that there is a family of adequate metrications of time based on our clock which are linear transformations of one another. The differences between the metrications reflect differences in the measure assigned to the duration of the clock events and in the interval assigned to the origin event.

It may be that there is some other physical system generating an isochronic sequence of events which could serve as our clock C_1. In which case the events generated by C_1 will last p/q as long as the events generated by the other clock, C_2, for some integers p and q. It does not matter which clock we select, for in either case we obtain the same family of adequate metrications. Any metrication based on clock C_1 will generate a family of adequate metrications which are linear transformations of each other. Given any metrication based on clock C_1 we can obtain a metrication based on clock C_2 if we multiply the measures of duration assigned by C_1 by p/q and correct the dates assigned through the use of C_1 by adding a constant. This reflects the fact that different events may be serving as the origins. That is, if an adequate metrication based on C_1 assigns to an end-point of an event the coordinate t, we obtain an adequate metrication based on C_2 if we multiply t by p/q and add a constant which equals the lapsed time between the beginning of the origin event in C_1 and the origin event in C_2. As this resulting metrication is a linear transformation of the original metrication it is already included in the family of adequate metrication. Similarly, from any metrication based on C_2 we can obtain an adequate metrication based on C_1 which is a linear transformation of C_2. Consequently, the choice between clocks whose events are related as indicated is an arbitrary matter for all c such clocks give rise to the same family of admissible metrications.

This account of the metrication of time brings into focus the high degree of idealization that is involved in the standard technique in physics of mathematically representing the topological and metrical aspects of time and things in time. The use of the real number system rather than the use of the rational number system is, as was argued in chapter VI, not empirically justified. It represents, if taken literally, the fictional supposition of a richness to time that there is no empirical reason to assume obtains. One might also object that assigning determinate intervals to events is another fictional imposition. For the temporal borders of events are vague. Consider, for example, a dog's barking. Is there either a last moment before the dog started barking or a first moment at which it is barking? If we represent the duration of the barking by an interval of the reals we are committed to thinking that one or other of these alternatives obtains. But what justifies us in making this assumption? One might hold that there is an interval of time before which it is false to say that the dog is barking and after which it is true to say that it is barking. But during the interval it is simply not determinate whether or not the dog is barking. Of course, once we have adopted a metrication for space and time we can describe events which will have determinate temporal boundaries. For instance, given a moving particle we can single out the event of its motion between the times t and t'. However, in assuming that all events including dog barkings are to be assigned intervals of real numbers we are being led by the precise character of our representational device to suppose without justification that all events have determinate temporal boundaries.

We have encountered a similar problem with regard to the relations such as 'lasts as long as'. For, as we noted, in direct empirical observation we do not find that one event lasts exactly as long as another. We find only that there is no discriminable difference between the events in regard to relative duration. Thus, in asserting that the events last exactly as long as, we go beyond the available data. One who wished to avoid such bold moves might seek a mathematical system which represented the relation of not being discriminably different with regard to duration. The difficulty is that this relation is not transitive and there are severe problems involved in finding a mathematical representation. For if we use a mathematical system which allows us to do what we want to do, namely assign a numerical value to a duration, we have to face the problem that the relation of being equal to, defined on the set of numerical values, is transitive. If we were to have an adequate mathematical representation we would need a mathematical system which was fuzzy in the way that the notion of not being detectably different in duration is.

3 CLOCKS AND CONVENTIONS

We have given a highly schematic and abstract account of a comprehensive system for assigning dates and measures of duration to any event given that we have some sequence of isochronic events produced by some 'clock'. This account needs to be filled out by outlining what considerations are relevant to selecting some actual physical system or systems to serve as a clock. The conditions that must be satisfied by any physical system which could serve as a clock will be outlined, and an account will be offered of the considerations that are relevant in deciding which of the systems that *could* serve as clock *should* be used as a clock.

This account is very much an idealization of our actual procedures for making temporal measurements. Among the assumptions that might not seem entirely realistic is the assumption that there is no event not covered by a sequence of clock events. If we were thinking, for instance, of our clock events as being generated by a pendulum clock this clearly does not hold. If we think that such a clock is the best to adopt, we will apply temporal measures to events not covered by means of some temporal property appropriately correlated with the pendulum clock in the range where both can be used. For instance, one might correlate carbon-14 decay with time as measured by the pendulum clock and extend the timing of events not covered by pendulum clock events using this correlation. However, it is the case that ideally we would like to adopt as a clock a type of system that does cover any event. This is in part at least why we prefer atomic clocks. For we think that any event is covered by some sequence of atomic vibrations.

Certainly it would not be appropriate to use some physical systems as clocks. Among those physical systems that we have occasion to employ as clocks some are, we feel, clearly better than others. We readily revise the judgments we make of the relative duration of a temporally separated pair of events on the basis of one clock system when presented with the results of comparing the same pair of events in some better clock system. It may seem to me that it took a lot longer today to walk to college and adopting my cheap and ancient pocket watch as the basis of a timing system may bear this out. If, however, when judged from the timing system provided by the telephone speaking clock my pocket watch seems to be wildly erratic in its behaviour, I would be ill-advised to regard my watch as providing a decent time system.

A clock, under the assumption that it generates a sequence of events of equal duration, or *isochronic* events, provides the basis of a system for making judgments about the temporal duration of other events. Without some clock system we have no basis, other than notoriously unreliably subjective impressions, of how much time has passed, for making such judgments. But we seem to have a vicious circle. If we need to appeal to a clock to justify judgments that, say, one event lasts as long as another, and if some physical system serves as a clock only if the events it produces are isochronic, how can we ever determine whether that assumption is warranted?

One answer to this dilemma runs, roughly, as follows. There are no grounds for claiming the assumption to be true. For any proposition of the form 'the events produced by the clock serving as the basis of a metrication are isochronic' is neither true nor false. It expresses a *convention* which we have adopted. Using that convention we can make judgments of temporal duration and those judgments can be said to be true or false *only relative to this convention*. Adopting some other convention (taking some other clock as generating isochronic events) might well require us to redistribute truth-values to our judgments of relative temporal duration. This position, which will be explicated more fully later in this chapter, will be called *Conventionalism.* Basically what the Conventionalist is claiming is that a clock is required if we are to make judgments of relative temporal duration and that it is *senseless* to ask of the events produced by this clock whether or not they are *really* isochronic. The events of the clock are *isochronic by fiat* and a search for the best clock is seen as a search for the convention that it is most convenient to adopt.

Conventionalism is not a natural view. It is unlikely to be adopted except on the basis of a philosophical argument. Pre-philosophically we are more likely to feel that it always does make sense to ask of any pair of events whether or not they have the same duration. Difficult or impossible as it may be to answer such questions we feel none the less that the events either did or did not have the same duration. This view, which will be characterized more precisely later in this chapter, will be called *Objectivism.* This is, roughly, the view that propositions concerning the relative temporal durations of events have objective truth-conditions that obtain or do not obtain independently of us. The Objectivist sees the task facing a would-be clock-maker as that of producing a physical system whose events *are* isochronic. The best clock is one which most closely approximates this condition.

The Objectivist and the Conventionalist need not disagree on the practical considerations to be taken into account in selecting a clock to serve as the basis of a metrication. The considerations which the Objectivist sees as guides to the truth, the Conventionalist sees as guides to the most convenient convention. For the Conventionalist changing clocks is a matter of opting for a new convention. For the Objectivist changing clocks is a matter of changing our minds about the truth.

4 CHOOSING A CLOCK

The disagreement between the Objectivist and the Conventionalist does not arise over the question of the factors to be taken into account in selecting a physical system to serve as a clock. It arises over the status to be accorded to the factors. Where the Objectivists see these factors as pointing to the truth of their claim that such and such a physical system really does produce isochronic events, Conventionalists see these factors as indicating the greater simplicity that will be obtained by stipulating that that physical system generates isochronic events. In isolating the factors involved, it will be fruitful to imagine that we are endeavouring to select *ab initio* a physical system to serve as a clock. This procedure is to be regarded as a heuristic device. In actual practice the decision to prefer one physical clock to another is made against a rich background of assumptions (many of which are highly theoretical) about the physical world, and a knowledge of the consequences of such choices that have been made in the past.

Beginning *ab initio* we have to reckon with the fact that we have a somewhat unreliable, but not all that unreliable, sense of the relative duration of events of which we have experience, which derives from what might be called our internal *biological clock*. For that reason we select for consideration as possible clocks only those physical systems that do not give results widely divergent with our intuitive sense of the passage of time. If this condition is not met in the situation we are envisaging the question will arise as to what grounds there would be for thinking of the physical system in question as measuring *time*. It should be noted that this very weak condition is defeasible. For instance, it would be abandoned if the only viable physics rested on a metrication which was at odds with our intuitive sense of the passage of time and if that physics offered an explanation for the divergence of our biological clocks from the clocks used in the physical theory.

If our clock is to be useful it must be readily available. This pragmatic factor governing our choice condition will be met if we base our metrication of either on a type of physical system which is easily reproducible, as mechanical clocks are, or on a type of physical system that can be observed widely, as the sun's rotation can be. Restricting attention for the moment to a reproducible system, we can see that this condition requires that different instances of the type of system in question keep in step with each other. Suppose a number of clocks are assembled, synchronized and then moved about to time various events. If they were found not to be synchronized when later brought together, we would have no basis for making judgments of the relative temporal duration of the events timed by these clocks. We require clocks which tend to keep in step, or as we shall say, tend to preserve *congruence* when subjected to variations in their physical location and conditions. It would be too restrictive to require that clocks keep perfect congruence. For given any type of physical system, it is likely that we can find some conditions which will destroy congruence. This defeasible condition means only that failure to preserve congruence should be the exception rather than the rule. In addition, when congruence fails it should be attributable, *prima facie*, to some obvious difference in the physical conditions to which the clocks have been subjected. If we were to think of adopting some non-reproducible system such as the rotation of the earth around the sun as providing our clock, we will need to supplement this by reproducible clocks. In this case we have a similar condition to that of congruence. For we require that the non-reproducible system be such that there are reproducible systems which tend to generate isochronic events when judged from the standard of the non-reproducible clock.

Any non-reproducible clock or reproducible system of physically similar clocks that tends to meet these conditions of congruence will be said to be a *prima facie* or *reasonable clock system, RCS*. In adopting an *RCS* we presuppose that these conditions are fulfilled. This empirical assumption can turn out to be false. Thus, both the Objectivist and the Conventionalist will agree that the selection of a clock involves empirical presuppositions, even though they disagree on the issue as to whether it is an empirical assumption that a clock generates isochronic events. For the fact that a family of physically similar clock systems tends to preserve congruence does not establish that each clock generates isochronic events.

A whole host of physical systems will pass the test of being *RCS*s — water clocks, hour glasses, digital watches, pendulums, atomic clocks and so on. The next stage in our rational reconstruction of the process of selecting a physical system to serve as the basis of a metrication is to elaborate the factors relevant to choosing between *RCS*s. This choice will be intimately connected with our physical theories in two related ways. First, it will depend on our physical theories in general. Second, it will depend on our particular theories covering the behaviour of the physical systems which constitute *RCS*s. Taking the general consideration first, let us suppose that we set about developing a theory of simple mechanics in the course of which we hit upon, say, Newtonian mechanics which includes the law that bodies not acted upon by forces continue at rest or in uniform motion. It will turn out that the empirical data gathered by, say, using water clocks will not fit this law as well as the data gathered using, say, digital watches. We could adopt water clocks as providing our standard for making judgments of the relative duration of events, in which case we would have to conclude that the Newtonian law was at best a fairly crude approximation to the truth. On the other hand, we could opt for digital clocks relative to which the data fits the law much better. Given that the theory in question is the best we have hit upon, there is no doubt that we will prefer the digital clock. Indeed, given the overwhelming success of the theory we are likely to be largely guided in our choice of clocks by the desire to arrive at data which fits the theory. We might even opt for the clock that gives perfect fit; namely, a body isolated from forces moving over equal spatial intervals. Thus, the selection of what I will call a *preferred clock system* or *PCS* is intimately bound up with the choice of physical theories.

An additional, related, factor can be isolated if we note that even if the clocks which are instances of a particular *RCS* tend to preserve congruence, there will be conditions in which congruence is destroyed. If a mechanical clock is subjected to extremes of heat and cold, tossed about severely and so on, it is likely to fail to keep time with a physically similar clock that is kept in 'normal' conditions. By studying the behaviour of physically similar clocks we hope to arrive at an account of the conditions under which congruence is destroyed. This may lead to correcting-laws which tell us how much to correct the readings of a clock in congruence-destroying conditions. One will have a further reason for adopting such correcting-laws if the data gathered by these clocks when corrected provides a better fit with our general physical theories than the uncorrected data. Consequently, it will count in favour of one *RCS* over another that the one admits of better correcting-laws than the other. Thus, we have a second condition relevant to the selection of a *PCS* from among the *RCS*s. The choice of a particular *PCS* will be further reinforced if using the *PCS* together with our physical theory provides a better explanation of the deviations of another

RCS than can be obtained if we take the other *RCS* and our physics and attempt to account for the deviations of the *PCS*.

5 CONVENTIONALISM

Both the Conventionalist and the Objectivist can agree that the factors enumerated above ought to govern our choice of a preferred clock system. However, the Conventionalist sees these factors as pointing merely to the most convenient choice to make. The satisfaction of such conditions is not regarded as providing evidence for the claim that the events produced by the preferred clock are truly isochronic. Unlike the Conventionalist, the Objectivist takes the satisfaction of these factors as evidence for the hypothesis that the preferred clock is generating isochronic events. In attempting to adjudicate between the philosophical perspective of Conventionalism and that of Objectivism it will be necessary to provide more adequate characterizations of these views than has yet been provided. To this end we will consider the version of the conventionalist thesis articulated by Reichenbach,[1] according to whom it only makes sense to ask of two events occurring at different times whether or not they have the same temporal duration *relative to a congruence standard*. The only pairs of events that can be compared without reference to a congruence standard are events occurring in the same spatial region which are such that either they begin and end together or are such that one begins after and ends before the other. Specifying the congruence standard for temporal duration involves, for Reichenbach, the following two elements. The '*first metrical co-ordinate definition*' provides the *unit of time* to be used by specifying some periodic process which is to serve as a clock and stipulating that a particular occurrence of that process marks one unit of time. The '*second metrical co-ordinate definition*' asserts that successive occurrences of the process in question are equal with regard to temporal duration and are to be assigned one time unit. The assertion that successive occurrences of this process are equal as to duration has, according to Reichenbach, the status of a *definition*.

> The equality of successive time intervals is not a matter of *knowledge* but a matter of *definition*. . . . All definitions are equally admissible.[2]

Reichenbach's grounds for thinking statements of this form to have definitional status seem to stem from his general view that if a meaningful statement is not testable it can only be meaningful in virtue of being either an analytic truth or a stipulated definition.

How can we test this assumption that the events generated by our clock are isochronic? There is only one answer: we cannot test it at all. There is basically no means to compare two successive periods of a clock, just as there is no means to compare two measuring rods when one lies behind the other. We cannot carry back the later time interval and place it next to the earlier one.[3]

Let us for the sake of argument assume that this assumption is not testable even by indirect means. What, then, is its status? It does not look like a stipulated definition. For the expressions 'being equal to with respect to duration' or 'being isochronic' already have a meaning. And it does not look like an analytic truth. For we take it, *prima facie* at least, that someone can have a full grasp of the sense of these expressions without realizing that the use of this expression is tied to any particular clock. If you and I are employing different physical systems as our clocks and if this leads you to say that E_1 lasts as long as E_2, and if this leads me to say that E_1 does not last as long as E_2, it does not look as though we have an equivocation of the meaning of 'lasts as long as'. Naively one wants to say that we agree on the sense or meaning but disagree on the extension. Reichenbach's description of my assertion that my favourite clock generates isochronic events as a stipulative definition is unfortunate. However, the basic Conventionalist thesis can be more happily characterized as the thesis that judgments of relative temporal duration lack truth-conditions. That is, the Conventionalist claims that there is no matter of fact at stake with regard to such judgments. There is nothing in virtue of which they can be true or false. The tenability of Conventionalism so characterized rests on two theses, one of which is empirical, the other philosophical.

The empirical thesis is the thesis of the *underdetermination of the metrication of time by the date* (hereafter cited as *UM*). The content of *UM* is given as follows. It is possible to produce a family of clocks which are pairwise non-linearly related to one another (as we saw in section 2 of this chapter, there is no essential difference between clocks that are linearly related to one another). Each clock has an associated total physical theory. No experiment or observation can decide between these pairs of clocks and theories. That is, no matter which choice we make we will not be thwarted by the outcome of any observation or experiment. The philosophical thesis is the thesis that there is no reason to assume that there is a matter of fact at stake if the supposition means that the matter of fact would be inaccessible; or, in other words, if a hypothesis is empirically undecidable even by the totality of all possible observations and experiments, there is no matter of fact at stake in virtue of which the hypothesis is true or false. This thesis (to be further discussed in section 7 of chapter X) will be called the *thesis of the essential accessibility of facts*, or *TEAC*.

The bold thesis I have attributed to the Conventionalist that *no* judgments of relative temporal congruence have truth-conditions is absurd. For if I adopt some deviant clock which gives the ice age, the time between my last two heartbeats and a performance of Wagner's Ring the same duration, I am just wrong. To say otherwise is to say something false and not merely to fix an eccentric extension for the notion of being isochronic. Some have thought that this licentiousness on the part of the Conventionalist provides a *reductio ad absurdum* of his position. However, if I am correct in taking it that what lies at the heart of Conventionalism is *UM* and *TEAC*, the Conventionalist can be more restrained if he holds that judgments of relative temporal duration have truth-conditions up to underdetermination. In explicating this constraint on Conventionalism we need to remember that the choice between clocks is intimately bound up with the choice of physical theories. If on the basis of adopting clock C we cannot arrive at a viable total physical theory, the assumption that that clock generates isochronic events should be rejected as false. Only if we have a clock C_1 and a total physical theory T_1, and a clock C_2 with an associated total physical theory T_2 incompatible with but empirically equivalent to T_1, is the choice between C_1 and C_2 undecidable. Consequently only then is there no grounds for thinking that one of these clocks is a true clock. Given *TEAC*, the Conventionalist will hold that in this case there is no fact of the matter at stake. In any case in which a clock does not give rise to a satisfactory total physical theory at all or gives rise to an empirically inadequate theory, the claim that the clock generates isochronic events can be rejected as false. And if we have reason to think that we have a unique total physical theory T_1 with an associated clock C_1, and no reason to think that there is any other empirically adequate total physical theory T_2 with clock C_2 non-linearly related to C_1, we are entitled to assert that it is true that the events generated by that clock are isochronic.

If Conventionalism is taken as I have characterized it, a rationale is provided for the assumption that we may have to resort to fiat or arbitrary choice in determining the extension of the temporal relations in question. For that is what will face us if *UM* obtains. And, given *TEAC*, we cannot regard our choice as a guess about the facts. For there is just nothing in virtue of which such a choice could be right or wrong. On my construal of Conventionalism, the extension of the relation of being isochronic is to be fixed by reference to our best physical theories. If there is a uniquely best total physical theory with an associated clock, the facts of the matter determine that that clock generates isochronic events and that any other clock non-linearly related to that clock does not generate isochronic events. If there is a plurality of empirically equivalent theories with associated clocks which are non-linearly related to one another, and if there are no other relevant evidential considerations relevant to the choice between the theories, there is, given *TEAC*, no content to the assumption that one or other of these clocks really generates isochronic events. Whether *UM* obtains is an empirical question that it would be exceedingly immodest to answer positively given the current state of knowledge. Grünbaum has endeavoured to render the claim plausible by developing two allegedly empirically equivalent forms of Newtonian dynamics having rival metrications which are non-linearly related to one another.[4] However, to establish this result with regard to a pair of total physical theories would be a Herculean activity (we do not even have a single total empirically adequate total physical theory), the success of which is far from obvious.

6 OBJECTIVISM

The Objectivist thesis is that there is a matter of fact at stake with regard to all judgments of the relative temporal duration of events. Such judgments are held to have objective truth-conditions which are either fulfilled or not. In this century Objectivism has had an exceedingly bad press indeed. Through the work of Poincaré,[5] Reichenbach[6] and Grünbaum,[7] Conventionalism has become the received view. But if *UM* does not obtain there is no serious threat to Objectivism. Conventionalists have tended to be positivistic or neo-positivistic in outlook and it may well be that the narrow positivist conception of what counts as evidence for a theory has inclined Conventionalists to assume too easily that *UM* holds. Putnam, taking a wider view of the factors relevant to theory choice, has confuted *UM*:

Let us try to formulate total science in such a way as to maximise internal and external coherence. By *internal coherence* I mean such matters as implicity, and agreement with intuition. By external coherence, I mean agreement with experiments checks. Grünbaum certainly has not *proved* that there are two such formulations of total science leading to two different metrics for physical space-time.[8]

Certainly Grünbaum has not shown that. And, as we have noted, to establish *UM* would be a daunting task. Equally, Putnam has not shown that *UM* does not obtain. But in view of my arguments that plausible candidates for the underdetermination of theory by data arise in the context of the topology of time, I am inclined to take seriously the possibility of *UM*. Thus, in so far as the Objectivist bases his case on the denial of *UM*, while accepting *TEAC*, the issue between the Objectivist and the Conventionalist must be regarded as open.

Some Objectivists have rejected *TEAC* while conceding or remaining neutral on *UM*. These Objectivists maintain that judgments of the relative temporal duration of events have truth-conditions which are either fulfilled or not fulfilled. If we are unable to recognize (even in principle) whether or not they are fulfilled this reflects adversely on our epistemological capacities and not on the claim that they are either fulfilled or are not fulfilled. As I will argue in favour of *TEAC* in chapter X this particular Objectivist strategy will not be explored further.

I have argued that, given *TEAC*, Conventionalism should be adopted if the evidence supports *UM*. Grünbaum's position is somewhat at odds with this. For, he argues, it is the absence of an intrinsic metric for time which would make Conventionalism more appropriate than Objectivism. Grünbaum's notion of what it would be for time to have an intrinsic metric can be explicated roughly as follows. An entity is internal to an interval of time if and only if the existence of the interval depends on the existence of the entity. A relation R is intrinsic to a time system if and only if the holding of R between a pair of intervals is entirely dependent on the entities internal to the intervals. The metric of time is intrinsic if and only if the equivalence relation of being isochronic is an intrinsic one.[9] It is held by Grünbaum that the only entities intrinsic to intervals are their instants. If time is discrete, Grünbaum argues, the metric is intrinsic for we can define the relation of being isochronic as follows. Interval I is isochronic to interval J if and only if I and J have the same number of instants. Since intervals can differ in the number of instants they contain, there will be intervals of different lengths. If time is *either* dense or continuous, Grünbaum argues, it lacks an intrinsic metric. If time is dense but not continuous any interval of time will have the structure of an interval of rational numbers and, consequently, the set of instants contained in or constituting *any* interval can be put in a one–one correspondence with the set of positive integers. If time is dense and continuous, *any* interval of time will contain a non-denumerably infinite set of instants. That is, the infinite set of instants in an interval cannot be put into a one–one correspondence with the positive integers. This set of instants can be put into a one–one correspondence with an interval of real numbers in this case.[10] Thus, in neither case (density or continuity) can the intervals themselves differ in an intrinsic way. According to Grünbaum, discrete time comes with its own natural, intrinsic metric but dense or continuous time has no such metric and we must impose one.

It is here that for Grünbaum conventionality enters in. We have to decide which metrication to impose through selecting a clock and there is no fact of the matter as to which clock, if any, is the correct or true one. But Grünbaum's stress on density and continuity as the source of conventionality is misplaced. For if we have an empirically adequate total physical theory based on a clock C, and if no other empirically adequate theory can be developed based on a clock non-linearly related to C, we would have adequate grounds for regarding the judgments of relative temporal duration based on clock C as being true. The metrication so derived would not be intrinsic to time in the sense that Grünbaum has given that notion. However, there would be no grounds for maintaining that there was no fact of the matter at stake with regard to the judgments in question. As we have noted, Grünbaum believes in *UM* and it is *UM* together with *TEAC* on which the Conventionalist thesis must rest in the case of dense or continuous time.

An Objectivist who rejected *TEAC* while accepting *UM* would maintain that in dense or continuous time, an interval I either lasts as long as an interval of time J or it does not, whether or not we can find this out. That is, he would argue that these intervals can differ with regard to duration notwithstanding the fact that the set of instants contained in the intervals is the same size. On this view, intervals simply can differ with regard to duration and do so independently of any convention or choice we might make as to the clock to be adopted. No doubt Grünbaum would reject such a move out of hand. And this reveals his implicit commitment to *TEAC*. Bearing in mind this commitment, let us consider his thesis that there is an intrinsic metric in the case of discrete time, which, according to him, is what makes Objectivism appropriate for discrete time.

If time is discrete we are certainly not going to be able to determine whether an interval I lasts as long as an interval J by counting the instants in these intervals. As in dense and continuous time, our only access to the comparison of intervals is through the events occurring during those intervals. It might be the case that in a discrete time world there would be two logically incompatible but empirically equivalent theories, one of which assigned the same number of instants to the intervals marked by events E_1 and E_2, and the other of which assigned a different number of instants to these intervals. In which case, one who embraced $TEAC$, as I have suggested Grünbaum does, could have no grounds for assuming that as a matter of fact either E_1 lasted as long as E_2 or that E_1 did not last as long as E_2. One who insists that there really is some determinate number of instants in these intervals will be embracing inaccessible facts. Thus, one who accepts $TEAC$ will have to adopt the Conventionalist thesis in the face of this underdetermination notwithstanding the fact that time is discrete. Thus, the discreteness of time is irrelevant to the issue of Conventionalism versus Objectivism.

If time were discrete and if we had a physical theory which enabled us to assign a determinate number of instants to any interval of time marked by an event, we could establish a particularly nice measure of the length of intervals. This is the measure given in section 2 of chapter VI which assigns as the magnitude of the duration the number of instants in the interval. However, this is not the only possible measure. We could, as Grünbaum himself notes, have a measure which assigned different values to successive intervals having the same number of instants.[11] That it is formally possible to define such measures in discrete time does not itself establish the appropriateness for discrete time of Conventionality. Only if one of these measures can be embedded in the best total physical theory is Objectivism appropriate. If the rival measures have rival associated physical theories which constitute cases of the underdetermination of theory by data, then, given $TEAC$, Conventionalism wins over Objectivism. Thus, we see once again that the central question in the debate in question is that of UM. Whether time is discrete, dense or continuous is just irrrelevant to this controversy.

7 PLATONISM AND OBJECTIVISM

Some who argue for Conventionalism do so under the explicit or implicit assumption of *UM*. Others have been attracted to Conventionalism because they regard the Objectivist as committed to the further unsatisfactory doctrine implicit in the following oft-quoted passage from the Scholium to the Definitions in Newton's *Principia*:

> Absolute, true, and mathematical time, of itself, and from its own nature, flows equably without relation to anything external and by another name is called duration: relative, apparent, and common time, is some sensible and external (whether accurate or unequable) measure of duration by the means of motion, which is commonly used instead of true time; such as an hour, a month, a year.[12]

Newton is frequently taken (unjustly perhaps, but doing historical justice to Newton is not our concern here) as maintaining that equality with respect to duration holds or fails to hold between any pair of intervals, *I* and *J*, entirely independently of what happens during those intervals. We could put this as the claim that the truth-conditions for judgments of congruence do not make essential reference to what happens during the intervals. Against Newton's view, construed in this way, one might seek to apply the argument of Grünbaum considered above. If time is a continuous system of instants existing independently of things in time there is nothing in virtue of which one interval could be the same length or a different length than another, for each interval is composed of a non-denumerably infinite set of instants. As such this set has the same cardinality as the set of instants comprising any other interval. Hence, unless the truth-condition for judgments of congruence of intervals ties those truth-conditions to what is going on in the intervals it is difficult to see in virtue of what the judgments will be true or false. This argument does not in fact show an incoherence in the view it attacks. A theological Newtonian willing to deny *TEAC* could maintain that time consists of a system of instants, which God has distinguished and just He knows, of an interval specified by a pair of points whether or not it is the same length as any other interval specified by a pair of points. That is, he regards each interval as containing the same cardinality of instants, but regards it as a brute feature of intervals that they bear determinate temporal relations of lasting as long as or of lasting longer than one another.

One who is both a Platonist and an Objectivist need not make the implausible move considered above of admitting these inaccessible metrical facts. For he could modify his position and allow that metrical relations hold between temporal items only derivatively in the sense that metrical interval *I* last as long as (longer than) interval *J* if and only if the sequence of events occurring during *I* lasts as long as (longer than) the sequence of events occurring during *J*. And, under the further assumption that physical theory uniquely determines the truth-value of such judgments about events, the facts in virtue of which these judgments are true or false will be accessible facts. It should be remembered that the Platonist is committed to the possibility of empty time. This means that given the line of argument being considered (that metrical relations hold between intervals only derivatively) metrical concepts will simply not be applicable to empty time. This seems to be the position of Swinburne who is both a Platonist and an Objectivist. For having argued that time is of logical necessity unbounded to the past and to the future, he says:

> when there are no physical objects, one cannot distinguish periods of time from each other. One could not mark for instance an end to any period of time which began with the final end of all physical objects, because there would be no event by which to mark an end to it. One could not distinguish an hour from a day in a period without objects.[13]

And Bruno seems to have been an early exponent of this view:

> It is certain that if there were no motion nor change, nothing could be called temporal; but there would still be the same one time for everything and one and the same duration called eternity. Indeed, time in the sense of the age of any particular thing would not exist. Therefore the existence of time in its particular kinds depends on motion.[14]

We have seen that a Platonist can be an Objectivist without embracing the implausible claim that metrical relations hold between intervals of time independently of what is going on during those intervals. Of course, a Platonist may be a Conventionalist. That is, he might hold that the time system necessarily exists and possesses the standard topology as a matter of necessity, but he could adopt *TEAC* and maintain that as *UM* holds there is no matter of fact at stake with regard to metrical judgments. Thus, Platonism is, in essence, neutral on the Objectivist–Conventionalist controversy concerning the metric of time.

Reductionism is also neutral on this issue. A reductionist who holds *TEAC* and *UM* will be committed to Conventionalism. A reductionist who holds *TEAC* and denies *UM* will be free to embrace an Objectivist construal of the status of judgments of the temporal congruence of events. Indeed, it seems possible that Leibniz had inclinations in this direction. Clarke objected to Leibniz:

> that time is not merely the order of things succeeding each other is evident; because the quantity of time may be greater or less, and yet that order continue the same.[15]

In answer Leibniz replied:

> I answer, that order also has its quantity; there is in it, that which goes before, and that which follows; there is distance or interval. Relative things have their quantity, as well as absolute ones. For instance, ratios or proportions in mathematics, have their quantity, and are measured by logarithms; and yet they are relations. And therefore though time and space consists in relations, yet they have their quantity.[16]

Given Leibniz's reductionism, the thesis that time has its quantity amounts to the claim that events have their quantity. There is no suggestion, in this passage at least, that the quantity of events is conventionally imposed on them. Given the verificationist strain that runs through the correspondence with Clarke, it would be anachronistic, but not unfair, to ascribe to Leibniz a belief in *TEAC* together with a disbelief in *UM*. Hence, there is a case for regarding Leibniz as a reductionist who holds an Objectivist position with regard to the metric of time.

Given the logical independence of the Objectivist–Conventionalist controversy and the reductionist–Platonist controversy it will not do to claim as Lucas does that:

> If we really regarded time simply as the measure of process, we should have no warrant for regarding some processes as regular and others as irregular. All would be equally good. Any continuous process could be made the standard by which we measured off isochronous intervals. It would be simply a matter of convention which one we adopted, not a matter of argument. But we do not regard it simply as a question of convenience whether the solar day, the mean solar day, or the mean sidereal day is really an isochronous interval or not. We are prepared to correct even the mean sidereal day, on

170

purely theoretical grounds and not for any considerations of practical convenience, to take into account the retardation of the tides on the earth's angular velocity; and the caesium clock is in principle subject to the same judgement — if, for instance, calculations in the General Theory of Relativity showed that frequencies were lower in gravitational fields. Even our best clocks are subject to correction. So long as we are prepared to assess the time-keeping qualities of a clock, and are prepared in principle to replace it by a more regular one, if it could be obtained, we are committed to an ideal of absolute time which is not simply what the clocks actually say.[17]

A reductionist (who in effect takes it that time is the measure of process) who held that there was a unique total physics which determined a particular metric is free to embrace Objectivism. Furthermore, a Conventionalist can assess the time-keeping qualities of his favourite clock without implicitly embracing Objectivism. For the Conventionalist such an assessment amounts to considering whether another clock might be more convenient. He can be prepared to shift to the more convenient clock without regarding such a shift as an attempt to obtain a clock which is more likely to be a 'true' clock. If the Conventionalist introduces a correcting factor to the readings generated by his actual clock he can talk of an ideal clock, meaning thereby a physical system which when used as a clock would give the same results as the actual clock after correction without regarding this 'ideal clock' as a *true* clock. Indeed, the sophisticated Conventionalist holds UM and hence maintains that there is a family of such 'ideal' clocks non-linearly related to one another, in which case the truth-conditions suggested by Lucas which refer to ideal clocks would lead to incoherence. That is, one of the ideal clocks will lead to judgments incompatible with the judgments based on other ideal clocks. Thus, we cannot coherently treat all the rival ideal clocks as giving the truth-conditions, and as there is, *ex hypothesi*, no way of determining which of these clocks is the true one, the sophisticated Conventionalist rejects the view that judgments of temporal congruence have truth-conditions.

8 OBJECTIVISM AND SEMANTIC REVISIONISM

The Objectivist has to convince us that any judgment of the form 'event A lasts as long as event B' is true or false in virtue of how the world is independently of any *non-trivial* choice, stipulation or convention on our part. That is, he must show us that judgments of this form have truth-conditions. The Conventionalist, since he accepts *TEAC*, will not be satisfied unless it is shown that the judgments of this form have truth-conditions which are such that we could in principle have evidence as to their satisfaction or non-satisfaction. Given *UM* the Objectivist cannot satisfy the Conventionalist's demands. The Objectivist may seek to refute *UM*. However, he can have no *a priori* guarantee of success. That is, even if he is able to show *de facto* that any rival pair of metrics are not on a par with regard to testability, he cannot have an *a priori* guarantee that this must be so. Alternatively, the Objectivist may concede the possibility of alternative metrications for time and deny *TEAC*, making what I called the ignorance response to *UM*. Interestingly, however, if the Objectivist makes this move the Conventionalist can side-step the whole thrust of the Objectivist's position without having to refute it. For the Conventionalist can simply say that he wishes to minimize, particularly in a scientific context, the number of predicates he uses that give rise to judgments whose truth-conditions cannot be recognized as obtaining or not obtaining. Consequently, he will stipulate by fiat that the predicate 'isochronic' does not have truth-conditions attached to its application. He will explain the conditions that count for or against advancing an assertion about temporal isochronity, and show how to work out the consequences of making one assertion rather than another, while refraining from claiming to have made anything true by fiat or stipulation. He sees his judgment that, say, atomic clocks generate isochronic events, as expressing his view about the best framework in terms of which to describe the world, and not as being true or false in virtue of how the world is. In making this move he cannot hold that the Objectivist's claim that the basic events generating the congruence standard either are or are not equal in duration is groundless. He has to say that he is operating in a language game in which that assertion has no role.

The Objectivist may argue that in the natural language, English, the sense given to sentences expressing comparative judgments of the form 'A lasts as long as B' is such that these sentences have truth-conditions. I have no idea whether such a case can be made out. But, even granting this, the Conventionalist can retort that this linguistic practice arose because of the false assumption on the part of the linguistic community that judgments of the form in question could be settled (in principle at least). That is, we have tended to see these judgments as objective because we have failed to see their undecidability. In any event, an interesting version of Conventionalism does not require a refutation of this claim about English. For the Conventionalist can offer his thesis as a reformative and not as a descriptive thesis. In so doing he is claiming it is preferable for scientific purposes at least to operate with a notion of being isochronic that does not give rise to judgments that are either true or false but which are undecidable.

A Conventionalist who regards his thesis as reformative rather than descriptive can point out that similar reformations have played an important role in progressive scientific change. For instance, one can see the Special Theory of Relativity in this light. Certain discoveries led to the realization that it is impossible to discover whether certain pairs of events are simultaneous. In the face of this realization, Einstein urged that we replace this notion of simultaneity by the notion of *simultaneity in a frame*. That is, in adopting the Special Theory we drop a notion of simultaneity that gives rise to undecidable judgments and replace it by a notion generating decidable judgments. We have here an illustration of an important but often unappreciated methodological principle, tacitly used in the construction of physical theories, to the effect that all things being equal, theories are preferable to the extent that what is said to be or not to be the case is restricted to what can be discovered to be the case or can be discovered not to be the case.

9 UNMEASURABLE TIME

I have considered what is required if Conventionalism is to be defensible either as a descriptive thesis or as a reformative thesis. It is easily seen that the claim advanced earlier in this book that time need not be measurable (in the sense that there are possible worlds in which no metrical relations hold between events) presupposes *TEAC*. For the Objectivist who denies *TEAC* can claim of any pair of events in any possible world that either they are or they are not temporally congruent. A world so chaotic that we could not in fact make any grounded judgments concerning the relative duration of events would be, for such an Objectivist, a world in which metrical relations held between events — metrical relations which we would be powerless to know. A Conventionalist or an Objectivist who holds *TEAC* will allow that if certain empirical conditions fail to obtain there are no metrical relations holding between events.

The empirical pre-conditions for the existence of measurable time that the Conventionalist has in mind are those conditions the satisfaction of which govern our choice of *RCS*. We can envisage a possible world in which there is no family of physical systems that keep even rough congruence and in which no corrective laws can be discovered that would establish congruence. In such a situation one could select a *particular* physical system and employ it in making judgments of temporal congruence. If there is nothing in favour of one such choice rather than another the measurement of time will be extremely arbitrary. It will also lose its point. We are interested in setting up systems for the precise measurement of time as this is an essential step in the process of discovering adequate scientific theories. However, if the world is such that no non-arbitrary metrical judgments can be made, it will be so irregular that we will not be able to discover any precise predictive scientific theories and will not have any use for precise metrical judgments. In the situation envisaged above, there is nothing in favour of selecting one particular physical system rather than another to employ in making metrical judgments. The choice is arbitrary not because any choice leads to an equally viable family of scientific theories but because no choice leads to any viable scientific theory. In such a world there would be no grounds for asserting that some event lasted as long as another and there would be no grounds for asserting that the one event did not last as long as the other. Consequently this world should be regarded as one in which the pre-conditions for measurable time are not satisfied. Events in this world would occur in an order but relations of comparative temporal duration would not obtain between them.

It remains to consider the connection between the topological aspects of time considered in chapter IV and the metrication of time. If time has a beginning and an ending one would intuitively expect the measure of all of time to be finite. Equally, one would expect that measure to be infinite if either time has no beginning nor end, or time has no beginning but has an end, or time has a beginning but no end. On the account of metricization that has been given that is just what we find. Suppose that time had a beginning and an ending. In this case time has the topology of a closed–closed interval of the reals (assuming for the sake of argument that time is continuous). As we have defined the measure of any closed–closed interval $[a,b]$ of the reals to be $|b - a|$, the measure of time will in this case be finite. Suppose that time has neither a beginning nor an end. In this case time has the topology of the real number line and the measure we have defined assigns an infinite value to any structure with this topology. And, similarly, if time has either an end but no beginning or a beginning but no end, the measure under consideration will assign an infinite value to the duration of time.

VIII

THE SPECIAL
THEORY OF RELATIVITY

Does Oxford stop at this train?
Apocryphally attributed to Einstein

1 THE SPECIAL THEORY OF RELATIVITY

No mention has yet been made of the Special Theory of Relativity (hereafter cited as *STR*) and as this theory is popularly supposed to embody a radical challenge to our ordinary concepts of space and time (and to the Platonist conception) this lacuna must now be filled. We need to explore the implications, if any, of the theory for the general debates between the Platonist and the reductionist and to explore what implications it has, if any, for the theses advanced concerning the topology and metric of time. To begin with, an account of the theory will be offered. There would be little excuse for offering yet another account of the theory if it were not for the fact that the most philosophically perspicuous way of developing the theory is as yet generally unappreciated by philosophers. This is the derivation of the Lorentz transformations by Zeeman.[1]

To begin with let us think in terms of our reconstruction of the notion of a timing system of using a clock to map events *in the vicinity of that clock* into some mathematical system in such a way as to generate a precise quantified representation of the dates and durations of these events. We will call such a date map a *local date map*. We are interested ultimately in assigning dates and durations to *all* events in the physical universe. To this end we can think of extending our local time map to cover all events. Such an extension will be called a *global date map*. If we assume a relation of simultaneity as an equivalence relation defined on the set of all events, we have a natural technique for extending any local time map. Any event is assigned the dates that would be assigned to an event simultaneous with it which occurs in the vicinity of the clock. If we wish to determine within the framework of classical physics the simultaneity classes, we could move a synchronized clock from the vicinity of the local clock to the events we are interested in. In this case we extend the local time map by making the assignments to non-local events that would be made by treating the transported clock as giving a local time map there. The transported clock would allow us, in classical physics, to determine the simultaneity equivalence classes. In effect, we arrive at a global time map in this case by patching together the local time maps of each spatial region.

As we noted in discussing the metric of time (ch. VII), all measuring systems embody empirical presuppositions which, if not satisfied, destroy the viability of the measuring system in question. The technique outlined above for generating a global extension of a local time map involves such presuppositions. For, suppose we found that a pair of synchronized clocks, one of which remains on earth, the other which is moved through space, fail in some cases to be synchronized when brought back together. Suppose further that the degree to which synchronization fails is a function of the path along which the moving clock is transported and of its speed relative to the clock at rest on the earth. Suppose, in addition, that we were unable to determine both a preferred speed and a preferred path. In this case there is no path and speed such that, if used, would give preferred assignments of dates and durations. That is to say, suppose, for example, we cannot explain by appeal to the presence of forces the relatively aberrant behaviour of all but one of the clocks. In this case we have failed to discover correcting-laws which bring all the transported clocks back into synchronization with the non-transported clock. If these possibilities were realized, as physicists claim they are realized, we could not use the method of clock transport to provide a viable base for extending local time maps into global time maps.

There remain, however, other alternatives. One might attempt to generate the partition of the set of all events into simultaneity equivalence classes by some method other than clock transport. If there were signals or causal chains that propagated with infinite velocity, we could use them to determine the equivalence classes. Events whose beginnings and endings could be linked by such signals or causal chains would be simultaneous. If there were no infinite causal chains but at the same time there was no finite limit to the speed of causal chains we could make our assignments of time to non-local events as accurate as we like by proceeding as follows. We record the departure time of a signal sent from our local clock to arrive simultaneously with the distant event. This signal is returned instantaneously with no change in speed and the time of its arrival at the local clock is recorded. Such a procedure is illustrated in the Minkowskian spacetime diagram below:

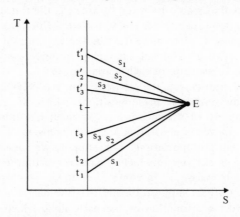

In this procedure we are assuming what I will call *the causality constraint*. The signal S_1 which arrives simultaneously with E was produced at the local clock at time t_1 and its reflection produced an effect at time t_1' at the local clock. Hence, the time assigned to its arrival at E must lie between t_1 and t_1', otherwise we violate the principle that causes come before their effects. By taking faster and faster signals (S_2, S_3, etc.) we can constrain as much as we like the time to be assigned to E. If it were the case that there is no finite limit to the speed of signals we can use this causality constraint to assign a date within whatever limits of accuracy we like. However, if we accept not only that there are no infinite signals, but also that there is a finite upper limit to *all velocities*, the speed of light (hereafter given as 'c'), this method will not give unlimited accuracy in the assignment of dates to non-local events.

Any rule which makes assignments of dates to distant events that satisfy this causality constraint has the following form: $t = \epsilon(t_1 + t_1') - t_1$, where ϵ is a parameter having values between 0 and 1. It is reasonable, on grounds of simplicity, to adopt what will be called the *first Einstein rule* by taking ϵ as ½. In this case $t = \frac{1}{2}(t_1' + t_1)$. This is tantamount to assuming that the velocity of the signal is the same in both directions. We need, in addition, a technique for assigning measures of spatial distance to non-local events which is given by $d(e) = \frac{1}{2}c\Delta t$ where $d(e)$ is the measure of the distance to e and Δt is the time required for a light signal to travel to e and back again. This rule, which will be called the *second Einstein rule*, is adopted for measuring distance in lieu of a rule in which distances were measured by transporting a rigid rod. This latter technique has to be rejected as the speed at which the rod is moved affects the result and there is no privileged speed at which a rod should be moved. Given that the first Einstein rule is to be used in assigning measures of distance, space and time are no longer entirely independent in their metrical aspects. For measures of distance depend on measures of time as $d(e) = \frac{1}{2}c\Delta t$. And the first rule makes the date assigned a distant event an implicit function of the spatial distance to that event.

I am treating this choice of a value for ϵ as a conventional choice. If one thought that the question as to whether or not light has the same velocity in either direction was open to empirical test one would not think of this as a conventional choice. However, one cannot put the question to any empirical test without first having a method for the synchronization of spatially separate clocks. We are forced to stipulate a value for ϵ here as we have no other method of synchronization.[2]

2 THE LORENTZ TRANSFORMATIONS

We have appealed to the causality constraint and considerations of simplicity in adopting the Einstein rules. This should not be allowed to disguise the fact that there is considerable implicit empirical content in the procedure. For these are viable rules only if there is an upper finite limit on all velocities. This is clearly an empirical fact that might have been otherwise. Interestingly, all that is required in order to pass from this slender basis to the Lorentz transformations which are the heart of the *STR* is a further causal postulate which is arguably a largely conceptual constraint. For if we assume a relation of causal connectibility defined over all events described in some global extension of a local time map and suppose that any other admissible frame of reference is one in which the extension of this relation of causal connectibility is preserved, we can show that any pair of frames are related by the Lorentz transformations. To this end we introduce the following definitions. Let M be some global spacetime map. A global spacetime map is a function which associates with each point in spacetime an ordered four tuple of real numbers (x,y,z,t) where (x,y,z) represent the spatial location of the point at the instant represented by t. For ease of exposition we will assume we are dealing with point-events rather than temporally extended events. Let e_1 and e_2 be events and let $M(e_1) = (x_1,x_2,x_3,t_1)$ and let $M(e_2) = (x_1',x_2',x_3',t_1')$. We define *the spacetime separation*, ds, between any pair of events e_1, e_2 as follows

$$ds^2 = (x_1' - x_1)^2 + (x_2' - x_2)^2 + (x_3' - x_3)^2 - (t_1' - t_1)^2.$$

Here we are assuming that the units of measurements of space and time have been so selected that $c = 1$. The expression

$$\sqrt{(x_1' - x_1)^2 + (x_2' - x_2)^2 + (x_3' - x_3)^2}$$

is the standard Pythagorean rule for determining the distance between the spatial location (x_1,x_2,x_3) of e_1 and the spatial location (x_1',x_2',x_3') of e_2. The expression $(t_1' - t_1)$ represents the time between e_1 and e_2. As we have set the speed of light at 1, the absolute value of the difference of t_1' and t_1, $|t_1' - t_1|$, represents the distance a light signal can travel between the time of e_1 and the time of e_2. If the spatial distance

$$\sqrt{(x_1' - x_1)^2 + (x_2' - x_2)^2 + (x_3' - x_3)^2}$$

between e_1 and e_2 is less than the distance $(t_1' - t_1)$ that a light signal can travel in the time between e_1 and e_2, a signal travelling less than the speed of light can connect e_1 and e_2. In which case

$$\sqrt{(x_1' - x_1)^2 + (x_2' - x_2)^2 + (x_3' - x_3)^2} < |t_1' - t_1|.$$

Since for any positive numbers a and b, $a < b$ if and only if $a^2 < b^2$,

$$\sqrt{(x_1' - x_1)^2 + (x_2' - x_2)^2 + (x_3' - x_3)^2} < |t_1' - t_1|$$

if and only if $(x_1' - x_1)^2 + (x_2' - x_2)^2 + (x_3' - x_3)^2 < (t_1' - t_1)^2$. Given the definition of ds^2 this condition obtains if and only if $ds^2 < 0$. We will say that events e_1 and e_2 are *causally connectible* if and only if

180

ds^2 is less than 0. This means in effect that two events are causally connectible if and only if a signal travelling with velocity less than the speed of light can link the two events. It is easily seen that this relation is a partial order. That is, it is reflexive, transitive and anti-symmetrical. A relation R is anti-symmetrical if and only if for all a, b, if Rab and Rba then $a = b$.

By appeal to a theorem of Zeeman's we can establish the following result. If we first define the set of *causal automorphisms* as the set of all maps, f, such that both f and its inverse preserve the partial order of causal connectibility, we can show that this set of transformations constitutes a group, the Minkowski group, which includes just the Lorentz transformations, reflections, rotations and multiplication by a scalar. No further assumptions are needed.

If we are dealing with a frame of reference F and a frame of reference F' moving with velocity v relative to F whose y and z axes are aligned, the Lorentz transformations are as given below, where the variables x, y, z, t give position in spacetime relative to F and the variables x', y', z', t' give position in spacetime relative to F'. Those on the left represent the transformations from co-ordinations given in terms of F to those given in terms of F' and those on the right are the inverse transformations from F' to F. As v' is the velocity of F relative to F', $v = -v'$.

$$x' = \frac{x - vt}{\sqrt{1 - \dfrac{v^2}{c^2}}} \qquad\qquad x = \frac{x' + v't'}{\sqrt{1 - \dfrac{v^2}{c^2}}}$$

$$y' = y \qquad\qquad y = y'$$
$$z' = z \qquad\qquad z = z'$$

$$t' = \frac{t - \dfrac{vx}{c^2}}{\sqrt{1 - \dfrac{v^2}{c^2}}} \qquad\qquad t = \frac{t' + \dfrac{v'x'}{c^2}}{\sqrt{1 - \dfrac{v^2}{c^2}}}$$

Writing β for

$$\frac{1}{\sqrt{1 - \dfrac{v^2}{c^2}}}$$

the transformations can be more tersely expressed as follows:

$$x' = \beta(x - vt) \qquad\qquad x = (x' - v't')$$
$$y' = y \qquad\qquad\qquad y = y'$$
$$z' = z \qquad\qquad\qquad z = z'$$
$$t' = \beta\left(t - \frac{vx}{c^2}\right) \qquad\qquad t = \beta\left(t' - \frac{v'x'}{c^2}\right)$$

The transformations corresponding to reflections deal with frames whose spatial axes are arranged as mirror images of one another as is illustrated for the two-dimensional case in diagram I below. Rotations deal with frames whose spatial axes are inclined at an angle to one another as in diagram II.

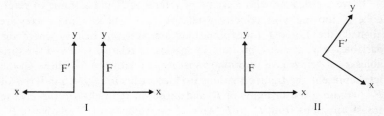

Multiplications by a scalar cover cases of frames in which different spatial or temporal units are used.

The postulate that we are using to derive the Lorentz transformation is the following: if the spacetime separation, ds^2, is less than zero for any pair of events described in one frame of reference it will be less than zero in any other frame of reference. The conceptual element in this postulate is the claim that if there is a relation of causal connectibility holding between a pair of events, that relation holds regardless of the frame of reference in terms of which the events are described. This is a conceptual and not an empirical assumption, for our notion of an objective public world involves the idea that the causal connections which do obtain do not depend on the spatio-temporal perspective from which the world is described. Certainly the postulate in question does have an empirical component. For it is an empirical fact that the expression ds^2 as defined does represent a relation of causal connectibility. Thus, we have the conceptual claim that causal connectibility is invariant and the empirical claim that a particular relation, ds^2, being less than zero, represents causal connectibility. The relation represented by ds^2 being less than zero is not intended to cover all possible causal connections but only those causal connections which propagate with velocities less than the speed of light. It should be added that the relation represented by ds^2 being equal to zero covers those causal connections that propagate with the speed of light. As Zeeman has shown, we can also derive the Lorentz transformations if we assume that this relation is an invariant.

3 CONSEQUENCES OF THE LORENTZ TRANSFORMATIONS

(i) *Relativization of Simultaneity and Platonism*

Given that all admissible frames are related by the Lorentz transformations, it follows that the simultaneity classes are not invariant from frame to frame. Let e_1 and e_2 be events assigned respectively the co-ordinates (x,y,z,t) and (x',y,z,t). In a frame moving at velocity v relative to the given frame, the times assigned to e_1 and e_2 respectively will be:

$$\beta\left(t - \frac{vx}{c^2}\right), \qquad \beta\left(t - \frac{vx'}{c^2}\right),$$

and given that $x \neq x'$,

$$\beta\left(t - \frac{vx}{c^2}\right) \neq \beta\left(t - \frac{vx'}{c^2}\right).$$

This might seem to establish, and is generally taken as establishing, an inadequacy in the Platonist conception of space and time. For the Platonist, thinking of time as a system of temporal items in which all events are located, has no room for the assignment within that system of different time co-ordinates to the same event. The Platonist must regard one or both of such assignments as mistaken. While this objection may make it uncomfortable to be a Platonist it is not telling. This result is uncomfortable as we tend to subscribe to the methodological principle cited earlier to the effect that all things being equal we should keep what is said to be or not to be the case as close as possible to what can, in fact, be known to be the case. Thus, if there are good scientific grounds for holding that we cannot ascertain, say, frame-invariant simultaneity classes it is preferable to revise the theory which leads us to posit such classes and to operate with a relativized notion of simultaneity which allows membership in these frame-dependent equivalence classes to be ascertained. This is not a telling objection to the Platonist view. For it reveals no vacuity or incoherence in the position. A militant Platonist can maintain that any event is or is not simultaneous with some other event. He will regard the reflections that led us to the *STR* as establishing only our *de facto* inability to discover these simultaneity classes. In doing so, he will be embracing the possibility of inaccessible facts. However, he may, like Swinburne,[3] hope for the triumph of a rival scientific theory which will restore knowable non-relativized equivalence classes and, hence, avoid having to admit inaccessible facts concerning simultaneity.

(ii) *STR and the Topology of Time*

From the representation of the *STR* given above, we can see that the theory has no implications for the issue of the topological structure of time. For relative to each frame of reference world history is mapped into the same four-space. Thus, whatever topological structure is assigned to time relative to one frame will be assigned to time relative to any other admissible frame of reference. The assignments of spatio-temporal location made in the differing frames are different. This is interpreted as reflecting the different relative locations and velocities of the frames in question. Indeed, the theory should be seen basically as a *metrical* theory. It is easily seen from the Lorentz transformations that while neither spatial separation itself nor temporal separation itself is frame-invariant, the spacetime separation given by the expression ds^2 is invariant. This means that against the background acceptance of the Einstein rules, the requirement that causal connectibility be invariant generates different assignments of spatial and temporal separation to the same pair of events in different frames.

The feeling that the *STR* has topological implications stems from a misconstrual of the fact that the order of all events is not invariant from frame to frame. However, the order of any events which are causally connectible either by a signal with velocity c or by an influence propagating with a velocity less than c is invariant from frame to frame. Only these events whose order within a frame can only be established metrically are assigned different orders in different frames. That is, no events except those to which an order can only be ascribed by first mapping those events into a time system (i.e. where, say, event E_1 is assigned time co-ordinate t_1 and event E_2 is assigned time co-ordinate t_2, and an order is ascribed to E_1 and E_2 on the basis of the relation between t_1 and t_2) have a frame-dependent order.

There is then no justification for the suggestion of Prior[4] that the *STR* requires linear proper time and branching public time. Prior considers the situation represented below where *b* and *c* are not causally connected:

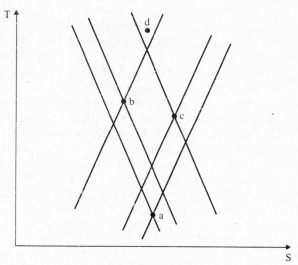

The interior of a forward light cone, say the one with vertex *a*, represents the set of all spacetime points which can be reached by a signal emanating from *a* with a velocity less than *c*. Spacetime points which can only be reached by a signal with the speed *c* lie on the lines defining the cones. Similarly the interior of the backward cone at *a* represents those spacetime points from which a signal travelling with a speed less than *c* could have reached *a*. Points on the boundary of the backward cone are those from which *a* could have been reached only by a signal with speed *c*. Given any forward light cone with vertex *a* and points *b* and *c* within that cone, *a* is causally connectible with *b* and with *c*. Hence *a* is before *b* and *a* is before *c* in all frames of reference since causal connectibility is an invariant. As light rays do not bend in the *STR*, there will be a point *d* in the forward cones of *b* and *c*. In all frames of reference *d* will be after *b* and after *c*. There is no possibility of causally connecting *b* and *c*. This means that in some frames *b* will be before *c*. In other frames this order will be reversed and in some frames *b* and *c* will be simultaneous.

Prior offers a set of non-metrical tense logic postulates for what he calls 'public or causal relativistic time' which give to this time a branching structure.[5] What led him to regard this time as branching is his assumption that while in all frames *b* and *c* have a common past, *a*, and a common future, *d*, there is no direct temporal relation holding in all frames between *b* and *c*. However, this is a highly misleading characterization of time in the *STR*. The *STR* does not assert that there is no temporal relation between *b* and *c*. What the theory asserts is that none of the relations being before, being after, or being simultaneous with, hold in all frames between *b* and *c*. If we are constructing a tense logic for the *STR* based only on invariant temporal relations, that logic must do justice to the fact that the following complex relation between *c* and *b* is, in fact, invariant. Either *b* is before *c* or *b* is simultaneous with *c* or *c* is after *b*. Thus, even restricting attention to public or causal time in the *STR*, this time is not branching. For there are invariant temporal relations holding between any pair of events and that is not so in the case of genuine branching time. Prior has attempted to develop a nonmetrical tense logic for the *STR*. Since the *STR* is basically a metrical theory, it is not surprising that his characterization is unsatisfactory. The moral to be drawn is not that the *STR* requires branching time but that time in the *STR* cannot be adequately characterized *via* a nonmetrical tense logic.[6]

(iii) *Quine and the STR*

It is not possible within the confines of this work to do more than note some consequences or supposed consequences of the *STR* that are particularly germane to the general discussion. Among these is the thesis advocated by Quine and others that the *STR* gives us no alternative but to treat time as being like space. If this were true it would be relevant to the claim advanced in the next chapter that our conception of time is a conception of an ordered structure with a direction. For space is not directed and if time is to be treated like space, time could not be directed. Quine argues for symmetry of space and time as follows:

> Just as forward and backward are distinguishable only relative to an orientation, so, according to Einstein's relativity principle, space and time are distinguishable only relative to a velocity. This discovery leaves no reasonable alternative to treating time as space-like.[7]

But the theory of relativity provides no support whatsoever for these contentions. The theory maintains an absolute distinction between space and time. This is reflected by the fact that the time co-ordinate in the semi-Euclidean Minkowskian metric has the opposite sign to the spatial co-ordinates:

$$ds^2 = dx^2 + dy^2 + dz^2 - dt^2.$$

It is easily seen that as ds^2 is invariant in all frames of reference, if there is a time-like separation between two events for some one observer, i.e., $dt^2 > 0$, and $dx^2 + dy^2 + dz^2 = 0$ then $ds^2 < 0$, and hence for all other observers $dt^2 > 0$.

In general the *STR* has no implications for how we ought to view the topological structure of time. Its implications are basically metrical in that it is arrived at as a consequence of having to abandon certain ways of extending local date maps to global date maps. The mode of this extension brings spatial parameters into play and leads to the non-invariance of spatial and temporal separation taken separately. However, the spacetime separation of events is invariant.

I have not been able to consider the range of alleged puzzles that some philosophers have thought to arise because of supposed tension between the *STR* and our ordinary conceptions of past, present and future. I am not convinced that there is any such tension. If there is such a tension I would argue that it is to be resolved through a modification of our ordinary conceptions of past, present, and future.[8]

4 THE TWINS PARADOX

One of the most intriguing aspects of the *STR* is the phenomenon of the *clock retardation*. Indeed, many regard it as paradoxical if not downright incoherent. Boldly and misleadingly put, the thesis which has given rise to this paradox-mongering is that moving clocks run slower. To see what gives rise to this thesis consider the situation illustrated where for the sake of simplicity we consider motion only in one spatial dimension. F_1 and F_2 are frames of reference with F_2 moving with velocity v relative to F_1 and F_1 moving with velocity v' relative to F_2. Let e_2 be a point-event occurring at F_1 whose temporal co-ordinate in F_1 is t_2. The spatial co-ordinate of e in F_1 will be 0. We assume that the clocks in F_1 and F_2 were synchronized and set at zero when F_1 and F_2 were in spatial proximity. In F_1, $t(e_2) = t_2$ and $x(e_2) = 0$. Let e' be an event occurring at F_2 whose temporal co-ordinate in F_1 is t_2. Since F_1 and F_2 were coincident when both clocks read

187

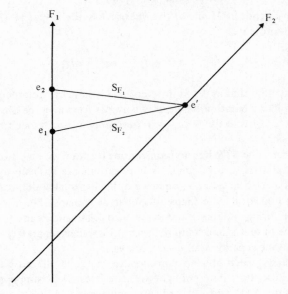

zero and since F_2 moves with velocity v relative to F_1, in F_1 the spatial co-ordinate of e' is vt_2. The line S_{F_1} in the diagram above joining e_2 and e' represents the fact that relative to F_1, e_2 and e' are simultaneous. To determine the time assigned in F_2 to e' we apply the Lorentz transformation as follows:

$$t(e') = \beta\left(t_2 - \frac{vx(e')}{c^2}\right)$$

$$x(e') = vt_2$$

$$\therefore \ t(e') = \beta\left(t_2 - \frac{v^2 t_2}{c^2}\right)$$

$$\therefore \ t(e') = \beta t_2\left(1 - \frac{v^2}{c^2}\right)$$

$$\beta = \frac{1}{\sqrt{1 - \frac{v^2}{c^2}}}$$

188

$$\therefore \ t(e') = \frac{t_2}{\beta}$$

$$\therefore \ \beta t(e') = t_2 \ \cdot$$

As c is the maximum speed of transmission, $v^2/c^2 < 1$. Hence $\beta > 1$ and $t(e') < t_2$. Thus, as F_1 assigns e' the time t_2 and F_2 assigns e' the time t_2/β, the clock in F_2 runs slow relative to the clock in F_1. It would be highly misleading to assert *simpliciter* that the clock in F_2 runs slow by the factor β. For all that has been established is that the clock in F_2 is slow *relative to the clock in* F_1. If we now consider the event e_1 at F_1 which F_2 regards as being simultaneous with e' we will see that the clock runs slow relative to the clock in F_2 by the same factor. Relative to F_2, $t(e_1) = t_2/\beta$ and $x(e_1) = v't_2/\beta$. To ascertain the time which F_1 assigns to e_1 we apply the inverse Lorentz transformation as follows. In F_1,

$$t(e_1) = \beta\left(\frac{t_2}{\beta} - \frac{v'x(e_1)}{c^2}\right)$$

$$x(e') = \frac{v't_2}{\beta}$$

$$\therefore \ t(e_1) = \beta\left(\frac{t_2}{\beta} - \frac{(v')^2}{c^2} \ \frac{t_2}{\beta}\right)$$

$$\therefore \ t(e_1) = \frac{\beta}{\beta} \ t_2 \left(1 - \frac{(v')^2}{c^2}\right)$$

Since the relative velocities of F_1 and F_2 are equal and opposite, $v' = -v$, and hence $(v')^2 = v^2$. Therefore,

$$t(e_1) = t_2 \left(1 - \frac{v_2}{c^2}\right)$$

$$\beta^2 = \frac{1}{\sqrt{1 - \frac{v^2}{c^2}}}$$

$$\therefore\; t(e_1) = \frac{t_2}{\beta^2}$$

$$\therefore\; \beta t(e_1) = \frac{t_2}{\beta}$$

Thus, F_2 assigns to e_1 the time co-ordinate t_2/β and F_1 assigns to e_1 the time co-ordinate t_2/β^2. As $\beta > 1$, $t_2/\beta^2 < t_2/\beta$ and, hence, relative to the clock in F_2 the clock in F_1 runs slow by the factor β. In the diagram above the line S_{F_2} represents the fact that relative to F_2 the events e_1 and e' are simultaneous. It is this difference in the planes of simultaneity relative to F_1 and relative to F_2 that displays the reciprocal relative retardation of the clocks.

We have seen how misleading it is to say boldly and without qualification that moving clocks are retarded in the *STR*. A clock C_1 moving relative to C_2 is retarded *relative to C_2*. And, reciprocally, the clock C_2 is retarded *relative to C_1*. Once it is noted that measures of retardation must be relativized to frames of reference it is easily seen that the *STR* does not, as Dingle fears, require:

> clocks to behave in an impossible manner for one set of synchronised clocks must concomitantly go both faster and slower than another set and this is impossible.[9]

Prima facie it seems that we can generate a paradox if we suppose that, say, digital clocks C_1 and C_2 associated with the frames of reference F_1 and F_2 respectively should be brought back into spatial coincidence. Surely it cannot be both that C_1 is retarded relative to C_2 and that C_2 is retarded relative to C_1. An examination of the clocks must reveal that one reads less than the other. They cannot both read less than the other. This is the essence of the infamous 'clock paradox', one anthropomorphic version of which runs as follows. You have a 'twin' travelling with F_2 who is biologically the same age as you when F_1 and F_2 are spatially coincident. The scare quotes around the word 'twin' are to indicate that he is not a genuine twin. God has fortuitously arranged things so that you are travelling with a non-accelerating frame F_1 whose path intersects with another non-accelerating frame F_2. The velocity of F_2 relative to F_1 is n. Travelling with F_2 is someone who is your physical counterpart at the time of coincidence of F_1 and F_2.

This benevolent but paradox-mongering God has arranged that a third non-accelerating frame F_3 intersects the paths of F_1 and F_2 (event E) as illustrated in the diagram. The velocity of F_3 relative to

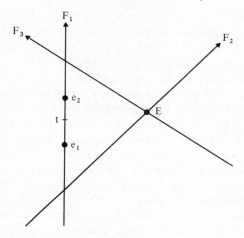

F_1 is $-v$. Happily, He has equipped each frame with a digital clock. The clocks C_1 and C_2 of F_1 and F_2 are synchronized and set to read zero when F_1 and F_2 are coincidental. The clock C_3 of F_3 is synchronized with C_2 when F_2 and F_3 are coincident and is set to read what C_2 then reads. Your 'twin' transfers frames when F_2 and F_3 coincide and transfers to your frame when F_1 and F_3 coincide. Let t be the time relative to F_1 when F_2 and F_3 are coincident and let $2t$ be the time relative to F_1 when F_3 and F_1 are coincident. Let T be the time relative to F_2 at which F_2 and F_3 are coincident. Since the situations with regard to F_2 and F_3 are symmetrical the time lapsed relative to F_3 between the coincidence of F_2 and F_3 (event E) and the coincidence of F_1 and F_3 will be T. Consequently, the total lapsed time for your 'twin' between your first meeting and your second meeting is for him $2T$. Applying the Lorentz transformation to obtain the time that he will assign to event E we obtain

$$T = \beta\left(t - \frac{vx}{c^2}\right)$$

$$x = vt$$

$$\therefore \ T = \beta\left(t - \frac{v^2 t}{c^2}\right)$$

$$\therefore \ T = \frac{t}{\beta} \ .$$

Thus, relative to your 'twin' the lapsed time is $2T$ which is $2t/\beta$. As $\beta > 1$

you regard your 'twin's' clock as retarded by the factor β. Thus, it looks as though more time has elapsed for you than for him. The appearance of a paradox is generated because you and your 'twin' agree that your parting from him and his parting from you are simultaneous. You also agree that your meeting him again and his meeting you again are simultaneous. Unlike the two-clock situation considered above, there does not appear to be the sort of disagreement about simultaneity that gave rise to the reciprocal relative retardation of the clocks. You are both comparing the time lapsed between E_1 and E_2, and the theory predicts that for you this is $2t$ and for him it is $2T$ where $2t = 2\beta T$. This looks like a non-reciprocal retardation that has worked to his advantage. Whether this represents a viable means of prolonging life given the assistance of a benevolent God raises biological and theological issues which fall outside the scope of this work. Consequently I propose to focus on the impersonal digital clocks. First, a word of explanation about the particular version of the so-called paradox that has been outlined. There has been considerable discussion of a version of the paradox in which one of two clocks initially at rest on earth is accelerated into space to be given later a deceleration and a reverse acceleration to bring it back to earth following a second deceleration. Applying the *STR* leads to the prediction that the travelling clock will be retarded relative to the clock remaining on earth. It has been held to be illegitimate to apply the theory to the accelerating clock to generate this prediction of non-reciprocal retardation on the grounds that the *STR* applies only to non-accelerating systems. This is just false. For the *STR* can be applied to the study of accelerating systems. Others who have accepted the legitimacy of applying the theory in this way have sought to explain the non-reciprocal retardation by appealing to the alleged force effects that the accelerations and decelerations will produce in the moving clock. A host of confusions lurk in this resolution which I have sought to avoid by focusing on three physical systems, no one of which accelerates.

In what follows I will introduce what I will call the *composite clock* C. This clock C is the clock C_1 between the coincidence of F_1 and F_2 and E and the clock C_2 between E and the coincidence of F_1 and F_3. Relative to C_1, the lapsed time between the departure of C and its return is $2t$. Relative to C, the time between departure and return is $2t/\beta$. I will assume that C in fact reads $2t/\beta$ on return. Does this mean we have in fact non-reciprocal retardation? If that were the case the lack of reciprocality cannot be explained by reference to the force effects during acceleration and deceleration as no physical system is either accelerated or decelerated. Interestingly it turns out that not-withstanding the fact that C_1 reads $2t$ and C reads $2t/\beta$ on reunion there is reciprocal retardation as the following argument reveals.

Relative to C_2 and relative to C_3, the event E has temporal co-ordinate T and spatial co-ordinate 0. Let e_1 be an event at F_1 which for C_2 is simultaneous with E. Thus, C_2 assigns e_1 the temporal co-ordinate T and a spatial co-ordinate, vT. Let e_2 be an event at F_2 which for C_3 is simultaneous with E. Thus, C_3 assigns e_2 the temporal co-ordinate T and the spatial co-ordinate $-v(-T)$ which is vT. C_2 whose velocity relative to C_1 is $+v$ applies the inverse Lorentz transformation to calculate the temporal co-ordinate of e_1 for C_1 and obtains the value

$$\beta\left(T - \frac{vx}{c^2}\,T\right)$$

which reduces to T/β. In a similar fashion C_3 whose velocity relative to C_1 is $-v$ applies the transformation to obtain

$$\beta\left(T + \frac{vx}{c^2}\,T\right) \qquad \text{or} \qquad \beta T\left(1 + \frac{v_2}{c^2}\right)$$

as the temporal co-ordinate of e_2 for C_1.

Consider now the composite clock C which undergoes the discontinuous change of motion from $+v$ to $-v$ at E. Taking one velocity (that of C_2), C regards the event (e_1) at C_1 which is simultaneous with E for C as having the temporal co-ordinate T/β. Taking the other velocity (that of C_3), C regards the event (e_2) of C_1 which is simultaneous with E for C as having the temporal co-ordinate $\beta T(1 + v^2/c^2)$. Since simultaneity *in a frame* is transitive and since, relative to C, e_1 is simultaneous with E and E is simultaneous with e_2, e_1 is simultaneous with e_2. But in C_1, e_1 has temporal co-ordinate T/β and e_2 has temporal co-ordinate $\beta T(1 + v^2/c^2)$. Consequently, C regards the clock C_1 as having jumped discontinuously from the first to the second of these readings. On meeting up with C_1 for the second time C argues that the reading on C_1 needs to be corrected by the factor $\beta T(1 + v^2/c^2) - T/\beta$ to take account of what he regards as C_1's great leap forward. Consequently, relative to C the time lapsed at C_1 is:

$$2t - \left(\beta T\left(1 + \frac{v^2}{c^2}\right) - \frac{T}{\beta}\right)$$

As $t = T\beta$ this becomes

$$2T\beta - \beta T\left(1 + \frac{v^2}{c^2}\right) + \frac{T}{\beta}$$

$$= T\left(2\beta - \beta - \beta\frac{v^2}{c^2} + \frac{1}{\beta}\right)$$

$$= \frac{T}{\beta}\left(\beta^2 - \beta^2\frac{v^2}{c^2} + 1\right)$$

$$= \frac{T}{\beta}\left(\beta^2\left(1 - \frac{v^2}{c^2}\right) + 1\right)$$

$$= \frac{2T}{\beta}.$$

C regards the time lapsed between the parting and the reunion as $2T$. Consequently, he regards the clock C_1 when corrected for its great leap forward as retarded by the factor β. C_1 regards the time lapsed as $2t$ and since the time lapsed for C is $2T$ which is $2t/\beta$, C_1 regards C as retarded by the factor β. Thus, notwithstanding the fact that C_1 has a greater reading than C on reunion we have reciprocal retardation. For while the actual reading on C_1 is β times the reading on C, the reading on C is β times the corrected reading on C_1.

North, in a careful study of the 'clock paradoxes',[10] considers a three-clock version akin to the one I have discussed. However, he fails to realize that there is, in fact, reciprocal retardation. This leads him to raise the question of the causal explanation of the retardation of C. Noting that an appeal to acceleration-induced force effects is inappropriate as no physical system accelerates, North suggests that anyone seeking a causal explanation will have to attempt to appeal to the relative motion together with the asymmetry in the paths of C_1 and C. However, if my account of the matter is correct, there is no non-reciprocal retardation to be explained. Obviously, there is an *apparently* non-reciprocal retardation but that is simply explained by the discontinuity of the motion of C which means that relative to C_1, C failed to record the passage of time between e_1 and e_2 because of the discontinuity in its motion. And this demonstrates why such composite clocks are unsatisfactory in the *STR*. For relative to such clocks a sequence of events, e_1 to e_2 in our examples, which are transparently not simultaneous are deemed simultaneous relative to the composite clock. Thus, I conclude that there is no clock paradox in the *STR*. The most plausible candidate for paradox – the three-clock system discussed – is easily shown to be non-paradoxical within the confines of the *STR*.

5 PLATONISM, REDUCTIONISM AND THE SPECIAL THEORY OF RELATIVITY

Almost invariably it is argued that the *STR* counts heavily if not decisively in favour of the Leibnizian-relational (reductionist) conception of space and time over the Newtonian-absolutist (Platonist) conception.[11] However, the situation is far from clear-cut and in this section I will argue that the *STR* can be developed either within a theory of spacetime that deserves to be regarded as an absolute theory or within a theory of spacetime that deserves to be regarded as a relational one.

As the account of the *STR* given above indicates, some major modifications need to be made in the classical conceptions of space and time. For classically space and time could be treated separately with the metric for space and the metric for time being independent of one another. In the case of the *STR*, space and time cannot be treated separately as the metrics for space and time are not independent of one another. This means that the theory is best developed in the context of a theory of spacetime for which a single metric given by $ds^2 = dx^2 + dy^2 + dz^2 - dt^2$. In order to give a perspicuous representation of the precise differences between classical and relativistic conceptions of space and time, it will be fruitful to characterize a classical theory of spacetime which can be generated by putting together the classical theories of space and time. To this end I give below a set theoretical characterization of what will be called Newtonian spacetime:

$\mathcal{N} = \langle P, S, G, T, t, d \rangle$ is a Newtonian spacetime if and only if

A_1 : P *is a four dimensional manifold of points.*
   P is to be thought of as the set of spacetime points.

A_2 : S *is an equivalence relation on* P.
   S is to be thought of as the relation of absolute simultaneity. For *p* in *P*, $[p]_S$ is the equivalence class of spacetime points under *S*. That is, $[p]_S$ is the set of all spacetime points simultaneous with *p*. These equivalence classes will be called *instants*. *P/S* denotes the partition of *P* by *S*. That is, *P/S* is

the set of equivalence classes under *S* (i.e., the set of instants of time).

A$_3$: G *is an equivalence relation on* P.

G is to be thought of as the relation of sameness of spatial location. These equivalence classes will be called *spatial locations*. *P/G* denotes the partition of *P* by *G* (i.e. the set of spatial locations).

A$_4$: T *is a linear continuous order on* P/S *having neither a first element nor a last element.*

T is to be thought of as the relation of being strictly before in time applied to instants. This has the effect of giving to time what I called in chapter III the standard topology.

A$_5$: t *is a real valued continuous function on the set of all intervals of instants.*

t is to be thought of as giving the duration between any two instants of time.

A$_6$: *d is a real valued continuous function on the set of all intervals of points of* P.

d is to be thought of as giving the spatial distance between any two points of the spacetime.

A$_7$: *Under the metric given by d each* P/S *is a three-dimensional Euclidean metric space.*

That is, each instant is a three-dimensional hyper-plane having a Euclidean structure.

A$_8$: For any equivalence class $[p]_G$ (i.e., any spatial location) there is for each *x* in $[p]_G$ a distinct equivalence class $[x]_G$ (i.e., an instant). For any equivalence class $[p]_S$ (i.e., any instant) there is for each *x* in $[p]_S$ a distinct equivalence class $[p]_G$ (i.e., a spatial location).

A$_9$: Se_1e_2 and Se_3e_4 and Ge_1e_4 and $Ge_2e_3 \rightarrow d(e_1,e_2) = d(e_3,e_4)$.

A$_{10}$: Se_1e_2 and $Se_3e_4 \rightarrow t(e_1,e_3) = t(e_2,e_4)$.

A simple model to illustrate a Newtonian spacetime is given below for a spacetime with one spatial dimension. Each point in the square represents a point of spacetime, the square being a sub-region of the spacetime. The horizontal lines represent some of the instants (the equivalence classes under S), and the vertical lines represent some of the spatial locations (the equivalence classes under G). World line L_1 represents a particle at rest; L_2 a uniformly moving particle and L_3 an accelerating particle.

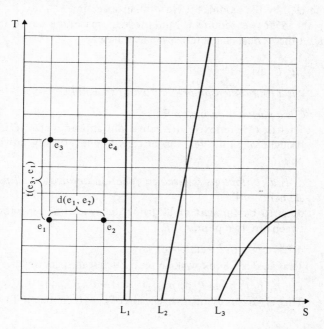

If one added to the above characterization the analogues of what I called in chapter I ontological and topological theses of the Platonist theory of time, this theory would be a Platonist theory of spacetime. The ontological thesis in this case would be the thesis that the manifold of points of spacetime exists independently of the contents of spacetime. The topological thesis would be the thesis that the spacetime has the structure attributed by the theory as a matter of necessity. In what follows I shall be concerned *only* with the theory obtained by adding to the axiomatic characterization given above the ontological thesis. On this theory, hereafter cited as *NST*, which has some affinities with Platonism, the question of the structure of spacetime is taken to be an empirical matter. The crucial feature of *NST* for my purposes will be its implicit denial of the claim that the points of spacetime can be treated reductionistically in terms of the contents of spacetime. By a

197

Leibnizian theory of spacetime, or *LST*, I will mean a theory that begins with the contents of spacetime, say, a four-dimensional manifold of *point-events* instead of a four-dimensional manifold of *spacetime points*, and construct from these ingredients the points of spacetime characterized by the axioms for Newtonian spacetime.

For the *STR* we require a Minkowskian spacetime which can be given a set-theoretical characterization as follows:

$$\mathscr{M} = \langle P, C, ds \rangle$$

M_1 : *P is a four-dimensional manifold of spacetime points.*

M_2 : *C is a partial ordering on* P.
That is, C is reflexive, transitive and antisymmetric. C is to be thought of as the relation of causal connectibility introduced on p.

M_3 : *ds is a continuous real-valued function defined on all pairs of elements of* P.
ds is to be thought of as giving the spacetime separation between any two points of P.

M_4 : ds^2 = dx^2 + dy^2 + dz^2 − dt^2.
That is, P is a Minkowskian semi-Euclidean spacetime.

M_5 : For p_1 in P, p_2 in P, d$s(p_1,p_2)^2 < 0 \leftrightarrow C(p_1,p_2)$.
That is, ds represents the relation of causal connectibility.

The theory of spacetime which is generated by adding to the above characterization the claim that the manifold of spacetime points exists independently of the contents of spacetime will be called the *Minkowskian absolute theory of spacetime*, or *MAST*. The reductionistic counterpart of *MAST*, to be called the *Minkowskian relational theory of spacetime*, or *MRST*, is the theory which starts with a manifold not of spacetime *points* but of spacetime *events* and constructs spacetime points out of these. Given that the proponents of *MAST* and *MRST* regard the investigation of the structure of spacetime as an empirical matter, the *STR* itself is entirely neutral between these rival theories. One who adopts *MAST* assumes there is a manifold of spacetime points whose existence is independent of the contents of spacetime. Consequently his theory allows him to admit the possibility of spacetime vacua and, indeed, to admit the possibility of an entirely empty spacetime. No more structure is attributed to spacetime than is required by the *STR*. And, in a sense, he attributes less structure to spacetime than the proponent of *NST* would. For instance, no relation of simultaneity is defined on the spacetime.

The advocate of *MRST* will have to deal with the problem generated by the possibility of vacua analagous to those which face the one who adopts a reductionist theory of time. In addition, an account must be provided of the method whereby the points of spacetime are to be constructed from the contents of spacetime. And, in doing this, the challenges similar to those facing the reductionist construal of temporal items noted in chapter VI will have to be met. The relativistic context complicates the required construction. As there is no frame-invariant notion of simultaneity, one cannot simply define instants as, say, equivalence classes of point-events. One has to define instead the notion of frame-bound instants as equivalence classes of point-events under the relation of being simultaneous in a frame. Of course, one inclined to *MRST* will argue that these challenges are not insurmountable and that they are a fair price to pay for the ontological economy of his theory. Interestingly, nothing in the debate between the proponents of *MRST* and of *MAST* will turn on the details of the *STR*, for both theories incorporate the *STR*. For this reason it is wrong to claim that the *STR* constitutes a vindication of relational approaches to space and time over absolutist approaches to space and time. Certain aspects of the classical absolutist conceptions of space and time are modified in the *STR* (i.e., there is no absolute simultaneity). However, these modifications must equally be made in classical relational conceptions of space and time. For instance, it is clear that Leibniz took simultaneity (defined on events) to be an absolute, frame-invariant, relation.Once the required modifications are made, we can elaborate two theories of spacetime, *MAST* and *MRST*, which differ markedly in their ontologies but do not differ at all with regard to the *STR*. *MAST* is in the tradition of the absolutist conceptions of space and time; *MRST* is in the tradition of the relationalist conceptions of space and time.

While there is a sense in which the *STR* is neutral with regard to the absolutist–relationalist controversy, the situation is not so clear with regard to the General Theory of Relativity (hereafter cited as *GTR*). For the field equations of *GTR*, as we noted in chapter II, admit of solutions in which the matter and radiation tensor is everywhere zero and the metric tensor is non-zero and, indeed, in some such cases the spacetime is curved. Thus, the vacuum solutions show that the theory *as it stands* is incompatible with any relational/reductionist programme for spacetime. For the theory represents empty spacetime as a physical possibility. Of course, one may argue 'that as a relational/reductionist programme is the best bet, one should assume that the field equations need modifying in some way to block the vacuum solutions. However, the results of the earlier chapters show that one cannot argue for constraining the theory field equations to block vacuum solutions on the grounds that empty spacetime is either incoherent or meaningless. Unfortunately, an exploration of the relevance of the *GTR* to the traditional absolutist–relationalist controversy falls outside the scope of this work. Elsewhere I will argue that the *GTR* requires a theory of spacetime which, in virtue of being anti-reductionist, has more affinities with the absolutist tradition than with the relationalist tradition.

IX

THE DIRECTION OF TIME

'Living backwards!' Alice repeated in great astonishment. 'I never heard of such a thing!'

'— but there's one great advantage in it, that one's memory works both ways.'

'I'm sure *mine* only works one way,' Alice remarked. 'I can't remember things before they happen.'

'It's a poor sort of memory that only works backwards,' the Queen remarked.

'What sort of things do *you* remember best?' Alice ventured to ask.

'Oh, things that happened the week after next,' the Queen replied in a careless tone.

Lewis Carroll, 1960, p. 172

I THE CRITERION OF TEMPORAL ORDER

With regard to a number of topological properties I argued — contrary to what has been often assumed — that it is an empirical matter whether or not time possesses these properties. In this chapter I shall extend this general view by arguing that it is an empirical matter whether or not time has a direction. This question of the directionality of time is a vexed, confused, and complex issue. And it will not be possible within the confines of this work to provide anything like a full treatment of the issues involved. There are, however, at least two reasons for including a preliminary discussion of the topic. First, part of the problem about the directionality of time is seeing what the problem is. I shall provide an account of what is the central problem. Second, I wish to show that there is a plausible view about the direction of time which is in keeping with the general tenor of the preceding chapters which deserves further exploration.

In what follows the term *directed order relation* will be used to refer to any two-place asymmetrical, transitive relation. The statement that time has a direction can be construed as the claim that it is appropriate in giving the order of things in time (and of temporal items) to employ a directed temporal ordering relation. I cannot here survey, for reasons of space, the differing and not always clear senses which various philosophers have attached to the expression 'the direction of time'. It does seem to me that the sense I have attached to the expression gives rise to an interesting version of the question as to whether or not time has a direction. I would emphasize that I do not claim to have made a fully adequate case for the theses advanced in this chapter. My intention is only to sketch a plausible line of argument concerning the direction of time which is in keeping with the general position maintained in this book.

If time necessarily has a direction it will not be possible to think coherently of a world of which it would not be appropriate to use a directed order relation in giving the temporal order of things in that world. If, on the other hand, it is a contingent matter whether or not time has a direction, there will be possible worlds of which it is appropriate to use a directed temporal ordering relation, and there will be possible worlds of which it is not appropriate to use such a relation. Certainly we do, as a matter of fact, employ directed ordering relations in giving the order of things in time in this world. For the terms 'before', 'after', 'earlier than', and 'later than' denote such relations. It will be fruitful to begin by considering the sense of terms for directed temporal relations which we regard as essential for expressing important facts about the world. This is easily seen if we note what would be left out of any description of the history of the world or a part of it which was given in terms of an undirected relation such as being temporally between. For instance, suppose that billiard balls are set in motion on a billiard table and that photos are taken at regular intervals of the table until the balls come to rest. Suppose you are given this set of photos ordered so as to represent the temporal betweenness of the states photographed. That is, you are given the sequence of photos $P1$, . . . , Pn but you are not told whether the process represented occurred in that order or in the reverse order $Pn, \ldots, P1$. Obviously something of importance has not been expressed in this mode of representation. The representation tells us only, for example, that the state represented by Pi came between the states represented by $Pi - 1$ and $Pi + 1$. Without the use of a directed ordering relation we cannot say whether the process happened in the order $Pi - 1, Pi, Pi + 1$ or $Pi + 1, Pi, Pi - 1$. For the purpose of the arguments of this chapter it will be adequate to single out for discussion the relation of being temporally before.

At one level we feel that our use of the relation of being temporally before is relatively unproblematic. For we can ostensibly indicate paradigm situations in which some event occurred before some other event. However, in view of the crucial importance of this relation for our entire network of temporal concepts we feel that some enlightening analysis ought to be provided. Here the matter becomes somewhat problematic. We might seek to give an analysis of this relation in terms of the notions of past, present, and future — as many writers have done.[1] However, this sort of analysis moves in too small a circle. For any attempt to analyse the notions of past, present and future brings us back to the notion of being temporally before. The classical attempt to give an analysis that breaks out of the circle of temporal terms involves an appeal to the notion of causality. Reductionists have been particularly interested in this sort of analysis. I characterized reductionism as the thesis that all assertions about time can be construed as assertions about temporal relations holding between things in time. The interest of some in the causal theory of time has stemmed from a desire to carry this reductionist programme a stage further by construing assertions about temporal relations holding between things in time in terms of assertions about non-temporal relations holding between things in time (the causal relation being taken to be a non-temporal relation). The causal theory of temporal order, which I shall argue is not satisfactory, asserts something like the following:

> An event A occurs before an event B if and only if either A is the cause of B or A is simultaneous with some event which is the cause of B or B is simultaneous with some event which is the effect of A or B is simultaneous with some effect of an event simultaneous with A.

There are a number of standard objections to this analysis which have been well canvassed in the literature. I note two of these without exploring them in detail. For I wish to raise a difficulty with this pattern

203

of analysis that will remain whether or not these particular objections can be answered. The articulation of the difficulty will lead us to the most important ingredient in the sense of 'before'.

First, one who adopts the causal style of analysis clearly commits himself to regarding the notion of backwards causation as incoherent, a view that must be regarded at least as contentious.[2] Second, if this is meant to be a *non-circular* analysis it must be backed by an analysis of the notion of causality that does not itself invoke the notion of temporal order. This of course means abandoning or modifying the classical Humean approach. However, my concern here is not with this putative link but with a more basic notion of temporal order, an understanding of which is presupposed in an understanding of causality.

That some notion of order is presupposed can be brought out by considering a third objection to the causal analysis. If it is given that there is a causal connection between events of type E_1 and events of type E_2, how do we ascertain which way the causal relation runs? Reichenbach, in particular, felt that the vindication of the causal approach required answering this question and attempted to do so through the use of his unsuccessful mark method.[3] But it is clear that there is a prior problem. How do we come to know that there is a causal link between events of type E_1 and events of type E_2 in order to pose the question of the direction of the link? Undeniably the chief source of evidence for the existence of causal links is the observation of constant conjunctions. In judging that there is a constant conjunction we must be employing some notion of temporal ordering. For we must have some notion which is involved in the judgment that instances of event type E_1 are to be grouped with instances of event type E_2, prior to asking whether E_1 causes E_2 or E_2 causes E_1 (assuming this is not an accidental correlation). Without such a notion we have no way of getting our experiential field organized into constant conjunctions at all. That is, without this, we might equally group all instances of event type E_1 together and group all E_2 type events together rather than forming pairs of events of type E_1 and type E_2. We can put the point as follows. Important as the appeal to causal connections may be in settling issues of temporal order, there must be some *epistemologically* prior notion of order which is employed in our coming to grasp that there are constant conjunctions or causal connections at all. The presence of a causal connection is not something given in experience. If we had, *per impossibile*, a direct perception of causal relations it might be that we could perceive temporal order through the perception of causal relations. We do not have such a faculty and hence we must have some notion of ordering that is implicitly or explicitly employed in warrantedly judging that causal relations obtain between events. If we employ some notion of temporal order in coming to make causal judgments the sense of our expressions for temporal order will be tied to the procedures we employ in making these pre-causal judgments of temporal order.

One might seek to justify this claim by appeal to the general principle that specifying the meaning of a predicate expression involves, in part, specifying the conditions which we take *paradigmatically* as warranting its application. This is a contentious principle whose justification falls outside the scope of the present essay. It is adequate for my argument to make the following weaker claim. Given a specification of the conditions which we *paradigmatically* take as warranting the application of a given predicate, the onus is on one who denies the general principle cited to show that this specification is *not* to be included within the specification of the sense of the predicate.Until it is shown that the specification in question is not to be included within the specification of the sense of the predicate it is reasonable to assume that it is to be included.

2 MEMORY AND TEMPORAL ORDER

Consequently, the task is to explicate some notion of temporal order that can be appealed to in vindicating claims of causal connection. Frequently we justify claims of the form, '*A* happened before *B*', by claiming to remember that *B* happened and to remember remembering that *A* had happened while experiencing the occurrence of *B*. This sort of appeal to memories cannot straightforwardly provide non-circular necessary and sufficient conditions of temporal order. For the mental states in question only count as states of memory as opposed to states of apparent memory if *A* really did happen before *B*. We cannot distinguish between memories and apparent memories here except by appeal beyond our mental states to the actual temporal relations between *A* and *B*. Nor can we provide in any simple way necessary and sufficient conditions in terms of memory impressions. For, notoriously, different persons not infrequently have incompatible memory impressions.

We cannot explicate in a non-circular manner the necessary and sufficient conditions for something happening before something else by reference either to memory or to memory impressions. It remains true that we do base our judgments of temporal order on our memory impressions. However, we demanded too much in demanding logically necessary and sufficient conditions for something to be before something else. If I am to place reliance on my current impressions as to what is now going on, I must place general reliance on my memory impression (at least for the reason that my confidence that I know the meanings of the words I am currently applying presupposes that memory impressions about past applications of these words by myself and others are basically correct). My memory impressions must be construed as providing presumptive evidence concerning what has gone on in the past and hence as providing presumptive evidence for temporal order in the following sense. Given memory impressions, what is required is not so much grounds for thinking them veracious but grounds for thinking them not veracious. What is required is not more evidence, which ultimately cannot be provided, but grounds for doubting the impressions. Grounds for doubting the veracity of memory impressions would include the sincere testimony of other observers who have differing memory impressions. The incompatibility of judgments of temporal order suggested by memory impressions with well-entrenched causal regularities would also provide such grounds. It is, of course, the reliance on the general veracity of memory impressions that leads to the positing of causal regularities which may subsequently be appealed to in testing the reliability of particular memory impressions.

The upshot of this is that there is some non-contingent link between temporal order and memory impressions. To say that A happened before B is to commit oneself to some loosely specifiable counterfactual about what memory impressions a *normal observer* who experiences both A and B would have, all things being equal. This means that if I assert A to have happened before B and deny that an observer, appropriately placed, would have the appropriate memory impressions, the onus is on me to give special grounds for thinking this. I may in some cases be able to do so by providing an account based on well-entrenched regularities which explain the aberration. It is not claimed that an explication of this link with memory impressions constitutes a full explication of the sense of our ordinary notion of temporal order. But it does at least capture an important component.

3 MEMORIES AND MORIES

Given that I am correct in claiming that there is a conceptual link between beforeness and memory, we can envisage a possible world to which we could not apply the notion of beforeness. In this section of this chapter I shall develop a characterization of such a world. In the following section the significance of this possibility will be considered.

To the first end imagine that we are possessed of a faculty analogous to memory but oriented towards the future. It is perhaps easiest to think of ourselves as gradually acquiring a faculty of precognition. We find ourselves gradually coming to have more and more impressions (with the same phenomenological vividness as memory impressions) about what is going to happen. These impressions, which we will call *mory impressions*, provide us with non-inferential knowledge of the future with the same general reliability that our memory impressions currently provide about the past. Doubtless the possession of such a faculty would require radical revision of our concepts and/or beliefs about human action. That such revisions would be required in no way shows that it is not a *contingent* fact that we do not possess such a faculty.[4] If we can distinguish memory impressions from mory impressions we can give a criterion for distinguishing whether A is before B as opposed to being after B as follows. Crudely, all things being equal, if while experiencing B we have memory impressions of A, A is before B. If, subject to the same provisos, while experiencing A we have memory impressions of B, A is after B. However, in the fantasy situation envisaged, we have both memory and mory impressions. We cannot distinguish between an impression as a mory impression rather than a memory impression on the grounds that one is an impression of something coming after the time at which the impression is had, while the other is an impression of something having come before the time at which the impression was had. For we wish to use the distinction between memory and mory to distinguish between before and after. If we use the term 'impression' to refer to either mory impressions or memory impressions we can express the situation as follows. While experiencing A we will have impressions of B and while experiencing B we will have impressions of A. Not being able to discern whether an impression is a mory impression or a memory impression we will not be able to attribute one order rather than another in the occurrence of A and B.

As is frequently said, time is our escape from contradiction. If we are, for instance, inclined to attribute to some persisting physical object, say a leaf, the properties of greenness and brownness, we avoid inconsistency by supposing it to have these incompatible properties at different times. So even in the fantasy situation being envisaged we will want to spread out in time the obtaining of states of affairs. As we must attribute some order to the ocurrences of which we have impressions, it is imperative to inquire how, in the absence of a viable distinction between before and after, we could do this. It would seem that if we imagine that the degree of detail and accuracy of our impressions at any time tend to fall off in either direction we might have a viable basis for employing a notion of temporal betweenness. For example, consider three events *A, B, C*. Suppose that while experiencing *B* our impressions of the occurrences at *A* and *C* are equally vivid and accurate. Suppose that while experiencing *C* our impressions of *A* are less detailed and accurate than our impressions of *B*. And suppose further that while experiencing *A* our impressions of *B* were sharper than our impressions of *C*. In such a case, this variation in degree of detail and accuracy could be used to place *B* between *A* and *C* in an undirected temporal ordering.

Our current ability to distinguish between before and after depends on our ignorance about the future. The ability of conscious beings in the fantasy situation to impute an undirected order of events would depend on their ignorance of things at a temporal distance in either direction from the present. It seems possible to conceive of conscious agents with a grasp of a notion of temporal order even if they are possessed with non-inferential knowledge of both past and future that covers equally all future and all past occurrences. For it might turn out that attributing one undirected order to their impressions led to a more reasonable, simpler, set of regularities governing all occurrences. A picture of the situation is of someone possessed with a set of pictures of a billiard ball at different positions on a table. By sorting these in one undirected order rather than another he is able to obtain a simpler characterization of the changes in the position of the billiard ball.

4 PHYSICAL CORRELATES

The picture of a world whose inhabitants have equally successful faculties of memory and mory might seem to be a picture of a world without directed time. For the inhabitants of such a world cannot apply our distinction between before and after and it would be very puzzling to maintain that their greater knowledge (in this case non-inferential knowledge of the future), relative to us, should bring with it an inability to discern some real feature of their world (in this case the difference between before and after). Puzzling as this might be, it does not follow from their inability to mark the distinction in question that the distinction is not there to be marked. To see this, consider a possible world whose inhabitants are unable to make colour discriminations but in which there are none the less coloured objects. We can bring out what must be added to the picture referred to above for it to constitute a picture of undirected time by considering further what is involved in the thought of a possible world containing coloured objects whose inhabitants are, in effect, literally colour-blind.

It is coherent to think of a world containing coloured objects whose colours are not perceived. We can provide grounds for the assertion that there are coloured objects in such a world by providing grounds for the following counterfactual. If *we* were present in that world we would be able to make the usual colour discriminations. If this counterfactual is true of the world in question that world contains coloured objects. If we were called upon to give some further characterization of a possible world whose inhabitants did not perceive colours which would provide grounds for asserting the counterfactuals, we could appeal to our general scientific theories of colour. We have well accepted theories of colour that explain why red things look red to us by appeal to the properties of light, the atomic structure of the objects, the nature of our visual apparatus and so on. By describing the world as one which satisfies the same laws as this world and as containing light, objects with the appropriate atomic structure, and inhabitants with appropriately deficient visual apparatus, we describe a world of which the former counterfactual is true. In thinking of a possible world to which some predicate does apply even though it cannot be applied by the inhabitants of that world it is essential to have a scientific theory concerning the application of that predicate.

I shall refer to phenomenal properties or relations which we apply to the world on the basis of our experience, and for which we have a physical theory of the sort we have for colours as properties or relations for which we have discovered a *physical correlate*. If we have discovered for a given property or relation a physical correlate we can derive from a description of an object given in terms of the physical correlate and the physical theory connecting the property or relation with its physical correlate, that the object has the property relation in question. Phenomenal properties or relations are those whose application we standardly apply to objects on the basis of the character of our experience. Redness is such a property, for to be red is to have a certain look in the appropriate circumstances. Unless we have a theory of the appropriate sort linking the phenomenal property P to the non-phenomenal property Q we have nothing to appeal to in warranting the assertion of a counterfactual of a possible world devoid of conscious beings of the form 'if anyone were to perceive object x in that world it would appear as P-type objects appear to us in this world'. However, if the object x had the property Q and there is a law-like connection between P and Q we will have grounds for asserting the counterfactual. For reasons of space this notion of a physical correlate has not been fully explicated. It does seem to me to be clear enough for my present rather minimal purpose. This is merely to sketch a line of argument which deserves further exploration and which, if it can be made out, would support the view that it is an empirical matter whether or not time has a direction. If we can discover a physical correlate for the relation of beforeness we can elucidate a picture of a world without directed temporality. Suppose for the sake of argument that we have discovered that whenever an event A is before an event B the total entropy of the universe at the time of B is higher than the total entropy of the universe at the time of A. Suppose further that this correlation is not a mere accidental correlation but is part of a well-developed, successful and highly confirmed scientific theory. In that case, a possible world in which entropy was constant and whose inhabitants had equally successful faculties of mory and memory would be a world without directed temporality. A possible world to which the entropy test gave a directed order and whose inhabitants failed to perceive any temporal direction would be a world with directed time whose inhabitants were 'blind' to the direction of time. A theory providing a physical correlate for the before relation might involve the claim that in the absence of directed time, consciousness could not arise. In this case our picture of a world without directed time will involve a world without conscious inhabitants, and not inhabitants who have equally successful faculties of memory and mory. I take it that we have discovered an appropriate physical correlation for the weaker undirected relation of temporal betweenness. Given a physical system of a certain type we can derive that some state S_2 is between states S_1 and S_3 by appeal to the values of the appropriate parameter at the times of S_1, S_2 and S_3, together with the laws governing that system. For example, if the system in question is a system of uniformly moving classical particles and if S_1, S_2 and S_3 are appropriate descriptions of the position of a particle, we can derive, in some cases at least, the result that occurred between S_1 and S_3 by appeal to the laws of classical mechanics.

The laws which provide the physical correlate for the relation of temporal betweenness are *time reversal invariant*. This means that we cannot derive from the description of two states of a system and from the laws governing that system which state came before which other state. And, consequently, the question of whether or not there is a physical correlate for the before relation is a matter of considerable controversy. Some theorists have sought to find a physical correlate in laws that are allegedly not time reversal invariant. Others have sought to locate the physical correlate not in the laws themselves but in the boundary conditions of the universe.[5] The question as to whether or not there is a physical correlate for the before relation is a question for empirical investigation. It is this question that seems to me to constitute the central problem concerning the direction of time. If an appropriate correlate can be found we can assert both that, as a matter of fact, time has a direction and that time might not have had a direction. It cannot be ruled out *a priori* that we will discover a physical correlate for the before relation of such a character that 'before' can be ascribed on the basis of the correlate only to events in our local region of spacetime. That is, the directedness of time might be a purely local phenomenon.

5 PREDICATES AND RELATIONS WITHOUT PHYSICAL CORRELATES

We tend to have a great deal of faith that a physical correlate can be found for the before relation. In part this faith is fostered by our past successes in finding physical correlates for properties and relations. However, there seems to be no reason to suppose *a priori* that such a correlate must exist. For instance, suppose that in spite of exhaustive scientific investigation we are unable to arrive at any satisfactory theory of taste. We are unable, let us suppose, to find a theory that would license the inference from the statement that some object possessed some property other than sweetness the statement that the object was sweet. As far as we can tell, in this case, there are no common properties other than sweetness possessed by all sweet things. Given repeated failures to find a correlate for sweetness it would be most reasonable to believe that sweetness lacked a physical correlate. In that event we would be compelled to regard the sweetness of things as arising, somewhat mysteriously, more from ourselves than from features of the objects which are said to be sweet. I shall say that in this case sweetness would be seen *not to be an objective feature* of objects which we describe as being sweet. By this description I intend the following. Given a possible world whose inhabitants did not make taste discriminations there would be no grounds for asserting counterfactuals about how objects in that world would taste to us. Hence, there would be no content in claiming that objects in that world were sweet. We would, of course, be justified in continuing to talk about the sweetness of some objects in our world. There would be, however, a good sense in which we could not regard sweetness as an objective feature of the objects of this world.

The various characterizations of the distinction between before and after as psychological, anthropomorphic or subjective that one frequently encounters[6] stem from a feeling that in the absence of a satisfactory theory positing a physical correlate of the before relation there may be no such correlate. *If* there is no physical correlate for the before relation there is a good sense in which it is not an objective matter whether or not time has a direction. In this case our attribution of a direction to time reflects something about our epistemological relation to the world and does not reflect something about the world itself apart from us. There is no content in the assertion that there would be distinction between before and after in *either* a possible world devoid of conscious agents *or* in a possible world whose inhabitants have equally viable faculties of memory and mory.

It seems, then, that whether or not beforeness has a physical correlate we can give a description of a world without directed time. If beforeness has a correlate such a world is described by specifying that the application of the test derived from the theory concerning the physical correlate fails to lead to a directed order. If beforeness lacks a physical correlate any world devoid of conscious agents with non-symmetrical faculties of memory and mory is a world relative to which there is no content in the claim that time has a direction. Furthermore, in the latter case, we have to regard our application of a directed order in giving the temporal order of things in our world as reflecting more a feature of our epistemological relation to the world than a feature of the world itself. Consequently, the central problem concerning the direction of time is the empirical question as to whether or not there is a physical correlate for the before relation. The status of our ascription of a direction to time depends on whether or not there is such a physical correlate.

X

TOWARDS A
POSITIVE THEORY

Time, like space, is not an ideal entity of a Platonic existence perceived by an act of vision, or a subjective form of order imposed upon the world by the human observer, as Kant believed. The human mind is capable of conceiving of different systems of time order. . . . Among this plurality of possible systems, the selection of the time order that holds for our world is an empirical matter. Time order formulates a general property of the universe in which we live; time is real in the sense space is real, and our knowledge of time is not *a priori*, but the result of observation. The determination of the actual structure of time is a chapter of physics — this is the result of the philosophy of time.
Reichenbach, 1959, p. 155

1 WHAT, THEN, IS TIME?

The arguments of the preceding chapters establish that the two theories of time with which we began, reductionism and Platonism, are untenable. One would like to be able to offer in their stead a positive general picture of time embodying an answer to the question — what is time? But, given what I called the promiscuous character of time, no such picture is going to contain all that it might be enlightening to say in answer to this question. However, one might aspire to provide a general picture of time which would unify, and perhaps in one sense explain, the particular conclusions we have reached concerning the topology and metric of time. And in the course of this chapter I will consider a number of such pictures. While several of these pictures are *prima facie* attractive they are ultimately untenable. We will see that it is their attractiveness that has beguiled many philosophers into thinking that the structure of time ought to be the subject of non-empirical, *a priori* investigation. In sections 2 and 3 of this chapter I will briefly reiterate the central objections to reductionism and Platonism. In section 4, van Fraassen's account of time as a logical space will be examined.[1] Further general pictures will be articulated in sections 5, 6, 7 and 8. In the end I will favour one particular picture — a plausible but neither a compelling nor an uncontentious picture.

In the discussion that follows I will appeal to three general constraints which, as the arguments of the preceding chapters show, must be satisfied by any satisfactory theory of time. First, any viable theory of time must display *at least some* assertions about time as having empirical content. I shall call this the *empirical constraint* and it will be taken as covering the following *two claims*. First, the existence of the time system is a contingent matter. Second, time possesses its topological properties contingently. The second constraint, to be called the *psychological constraint*,[2] is the following. Any theory of time must display time, or at least certain aspects of time, as not being mind-dependent. For we think that there is no incoherence in supposing the existence of a changing world in time which is devoid of conscious beings. The third constraint requires that the philosophical account of time be compatible with currently acceptable scientific theories. All things being equal, it should also be compatible with what we take to be promising lines of scientific enquiry. That is, given a good or promising theory, we must either find a construal of that theory which is compatible with our philosophical account of time or we must seek to modify that philosophical account to render it compatible with the scientific theory. I will call this constraint the *scientific constraint*.

2 PLATONISM

Platonism was construed as the thesis that time is a system of temporal items whose existence and properties are independent of the existence and character of the set of events constituting the history system. The arguments of chapters II, III and IV provide sufficient grounds for rejecting such a view. For we have seen that time admits of different topologies and, hence, any theory of time that entails the claim that time possesses its properties as a matter of necessity must be rejected by *reductio ad absurdum*. Thus, Platonism violates what I called the empirical constraint.

An account should be provided of the powerful attraction that this picture of time has had for some. First it may be that we are tempted to think of time as necessarily existing because when we imagine a possible world we imagine a world in time. This tempts us to think that *any* possible world is a world in time. However, we are overlooking the possibility of an empty world, a world devoid of empirical objects and hence devoid of change. Such a world would not be a world with spatial and temporal aspects. The existence of time is, almost, but not quite, a matter of logical necessity. For most, but not all, logically possible worlds are worlds embedded in a time system.

Second, we can introduce a name, *t*, for a time in this world by means of a demonstrative (i.e., the time *now*) or by means of a description (i.e., the time of such and such an event), and having so introduced the name '*t*' we tend to think that we have latched on to some item, *t*, the existence of which is independent of the event cited in identifying it. This is partly correct. We can think counterfactually of *t* as a time at which no one in fact said 'now' or at which the event in question did not occur. The identity of *t* is not necessarily tied to the particular event used to pick it out. We can sensibly suppose that some event other than the event *E* occurred at the time in question. Consider the time period marked by my cycling to College today. If my alarm clock had not gone off, I would have been sleeping during that period of time. There is a possible world quite like the actual world in which the alarm did not go off and there is a time which can be identified in that world as the same time period as the period in this world during which I cycled. This counterfactual supposition presupposes sufficient similarity between the two worlds to allow this cross-identification of times. In many cases such counterfactuals are entirely intelligible. However, if we think of a possible world radically unlike the actual world there may be no time in that world which is the same as the time of my cycling to College this morning. While I may be able to identify a time in some similar possible worlds as the same time as some time identified in this possible world there is not in *all* possible worlds a time which is the same as this time. Consequently we should not be led to think that we have identified some time, *t*, whose existence is entirely independent of events occurring at that time.

For the Platonist, if I succeed in identifying a temporal item in this

216

world that temporal item occurs in all other possible worlds. I may not be able to say which temporal item in some other world is that temporal item, but there is, none the less, one that is identical to it. By contrast, on the reductionist theory any possible world lacking a set of events which is the counterpart of the set of events constituting the equivalence class of some event cited in identifying a temporal item in this world, is a world in which that temporal item does not occur. The strength of the Platonist picture stems from the fact that I usually can identify a temporal item by citing some occurrence, and at the same time speculate about what would have characterized that temporal item even if the occurrence cited had not in fact occurred. The Platonist position overreaches itself in drawing the conclusion that any temporal item in any one world occurs in any other world. As there could be no way of carrying out this identification of times across possible worlds in all cases we find the assumption that there is such cross-identity untenable. The strength of the reductionist position stems from its apparent explanation of why this is unintelligible. For the reductionist, a time is just a set of events and there is no coherent notion of a time independent of the set of events.

3 TIME AS A CONSTRUCTION FROM EVENTS

We have seen from the argument of chapter II that this reductionist picture of time must be rejected. But I noted earlier (p. 47) that the thesis of *modal reductionism* cannot be rejected on the basis of the argument of chapter II. Modal reductionism does, however, face the difficulty noted in chapter III. We cannot regard all assertions about time as analysable into assertions about the actual temporal relations between actual things in time. It remains possible to adopt a form of *methodological reductionism*. In this case it is not claimed that reductionism embodies an adequate *analysis* of our concept of time. Instead it is claimed that one ought to keep as close as possible in what one says about time to what would be legitimate on reductionistic assumptions. Thus, for instance, one would try to avoid framing scientific theories which involved positing time without change. However, it could not be claimed that such theories were incoherent or vacuous. It is perhaps possible to interpret Mach as holding this methodological reductionist position. In discussing Newtonian conceptions of time he says:

With just as little justice, also, may we speak of an 'absolute time' — of time independent of change. This absolute time can be measured by comparison with no motion; it has therefore neither a practical nor a scientific value; and no one is justified in saying he knows aught about it. It is an idle metaphysical conception.

The stress here seems to be uselessness and not meaninglessness.[3] However, this methodological reductionism while perhaps defensible, *qua* methodological programme, provides no answer to our question: what is time? And, further, the arguments of chapter II display a context in which maintaining this programme would be decidedly *ad hoc*.

4 TIME AS A LOGICAL SPACE

Van Fraassen has argued that the best answer to the question — what is time? — is that it is a *logical space*. Van Fraassen, borrowing the general notion of a logical space from the early Wittgenstein,[4] articulates this view as follows:

> the position we are presently discussing is that time is a logical space and that a logical space is, in general, a mathematical construct used to represent conceptual interconnections among a family of properties and relations — and furthermore, that this logical space (time) is the real line being used to represent all possible temporal relations among events and the connectual interconnections among their relations.[5]

On van Fraassen's construal of this view, there are two essential ingredients. One ingredient is the notion of time as a *logical space*. The other ingredient is the notion of its being a logical space *used by us to represent* relations between events and conceptual connections between concepts. To take the latter aspect first, we certainly do use mathematical structures in representing the sort of thing that van Fraassen has in mind. However, the *use by us* of a certain structure as a representational device is not a necessary condition of the existence of time. For even if there had not been any representors there would have been time to be represented. Van Fraassen maintains, to the contrary, that there would be no time if there were no conscious beings, but he attempts to dilute this conclusion somewhat by arguing:

> A logical space is a mathematical construct *used to represent* . . . and

218

that means, of course, *used by us*. If we users and representors did not exist, neither would there be something being used to represent. The real line cannot be used to represent the actual temporal structures of events unless the latter can be embedded in it. This is purely and entirely an objective question of empirical fact. But neither can the real line thus be used unless there are those capable of using it. Hence in that case the logical space *time* (which is something used to represent something else) could not then exist.

But in this sense in which there would be no time if there were no beings capable of reason, is innocuous. It is the sense in which there would be no food were there no organisms, and no teacups if there were no tea drinkers. There could be things that look like what, in our world, teacups look like. There could be things that could be used to drink tea from (buckets, shells and so on). But teacups are the things that *we* use to drink tea, and in that sense they are as much a cultural object as chess or the Polonaise.[6]

This just does not reflect the content of our concept of time. *Arguably* an object is a teacup only if it has been produced by a conscious agent with some general intention about its use or possible use and, consequently, in a world devoid of conscious intending agents while there might be things *like* teacups there would be no teacups. However, this just does not seem to apply in the case of time. Our conception of time is not such that we would say of a possible world devoid of conscious agents that while there is something *like* time, there is literally no time on the grounds that no one was doing any representing in that world.

Notwithstanding the above objection, one can be sympathetic to van Fraassen's claim if it is understood as follows. One might wish to hold that a full grasp of our concept of time involves grasping the possibility of using certain mathematical structures in representing the temporal aspect of things. Someone who fails to see this has failed to grasp something about our *full* concept of time. This is part of our conception of time in contrast to a more primitive notion of time one can envisage, the possessors of which would not regard representability of this sort as part of their conception of time. But to hold that our conception of time is such that time is necessarily representable by one of a class of mathematical structures is not to concede that time as such *is* a structure *used* to represent. Hence one need not concede the mind-dependence of time.

Ignoring van Fraassen's stress on time as related to our use of certain representational devices one might hold that it is enlightening to characterize time simply as a logical space. If we attempt to construe this along Wittgensteinian lines the following problem arises. Wittgenstein clearly held time along with space and colour to be a logical space.[7] Just what Wittgenstein intended by the notion of a logical space is not entirely clear. However, it does seem fairly clear that a given logical space has the properties it does have as a matter of logical necessity.[8] For Wittgenstein, the only kind of necessity is logical necessity. The impossibility, for example, of two colours being simultaneously present at the same place in the visual field stems from the *logical structure* of colour. Presumably then, the logical structure of time entails that time *qua* logical space has a certain structure as a matter of logical necessity. And certainly, as was noted on p. 96, Wittgenstein held in his *Philosophical Remarks* that the non-beginning and non-ending did not reflect a matter of contingent fact. He held, on the contrary, that this reflected only a feature of the grammar of 'time'. While there are some properties of time that one might hold to be of this character, say, that events occur in time, the topological characteristics we have been considering have been argued to be properties the possession of which by time is a contingent matter.

Van Fraassen accepts that calling time a logical space means according necessity to time's possession of its properties. His strategy is to attempt to mute this conclusion by holding that the necessity in question is not logical necessity but is *necessity relative to a conceptual framework*. And, in turn, this seems to be construed as 'necessity relativized to the scientific theories that we accept'.[9] If we take it that on this notion of relative necessity a proposition, *P*, is necessary relative to theories, *T*, if and only if the conjunction of *T* and the denial of *P* are logically inconsistent, the claim seems trivial. For an adequate statement of any reasonably sophisticated scientific theory will contain propositions ascribing to time a certain topological structure. And, quite trivially, the conjunction of propositions specifying the theory and the denial of the propositions in question concerning the topology of time will be logically inconsistent. In this case no real content has been given to the claim that time necessarily possesses its properties. For under the above construal this amounts to no more than an assertion that time is said in certain theories to have a certain structure.

There is one further puzzling feature in van Fraassen's characterization of time as a logical space. He continually stresses that the notion of time as a logical space involves the notion of *time's being a mathematical construct used to represent.* This just seems confused. One does not want to say that the time system is just some mathematical object. One may want to describe the time system as having a structure isomorphic to some mathematical structure, for it is that isomorphism that entitles us to use the mathematical structure to represent time. But time is not the structure that is used to represent it.

5 TIME AS A SYSTEM OF NON-EMPIRICAL ABSTRACT OBJECTS

The properties of time cannot be discovered by *a priori* reflection. Many philosophers have taken the contrary view and it would not be satisfactory to conclude that they were simply mistaken without explaining what it is about time that allows their position to seem not entirely implausible. Consequently, I shall articulate in this section a picture of time in terms of which their position is not implausible.

At one time it was commonly thought that the properties of space could be discovered by *a priori* reflection. Given that the only known geometry was Euclidean geometry and given a belief that some geometrical characterization of space must be possible, it was not surprising that many thought that space *had* to have a Euclidean structure. For there were no known alternative structures. If one were in a position of having no representation of the logic of tenses other than one ascribing to time the standard topology, and if one did not have available the various set-theoretical resources which we have employed in characterizing different topologies for time, one might be inclined to think that time *must* have the standard topology. One might not be able to see in such a situation how it could be otherwise. For one would lack the resources to express adequately the thought that time might not have the standard structure. This consideration may have played a role in the past in influencing some to think that the properties of time can be discovered by *a priori* reflection.

As we noted in chapter I, we talk both about time and the history system. We talk as if time was distinct from the history system. We use the substantive 'time' in such a way as to suggest that time is that in which the history system is located. In so thinking, different possible worlds are treated as located in *time* and are not treated as each having its own time. Treating the time system as common to any possible world may generate a feeling that there must be things that can be established about time independently of the particular features of any given world. Time becomes a sort of container into which different worlds could be placed and its properties are not a function of its contents.

This crude picture of time as a container whose properties can be investigated without reference to its contents has, as a counterpart, the following Platonist picture. Assertions about the topology of time are true or false neither in virtue of facts about the history system nor in virtue of facts about the system of temporal items. Such assertions reflect the character of the conceptual framework for talking about the temporal aspects of world history. From an explication of the content of our temporal concepts one can derive the structure of time. One might think here of the certain views of numbers according to which substantive truths concerning numbers are discoverable by *a priori* reflection on numerical concepts. The time system is, on this picture, taken to be a system of abstract non-empirical objects. That is, neither the existence nor the properties of the time system depend on the empirical world. Hence the structure of time is open to *a priori* investigation.

It has already been argued that this Platonist conception of time has to be rejected. My intention in reiterating this characterization of it is not to develop arguments in its favour but to explain why the view has had the currency it has had. My suggestion is that if one is inclined to view time as a system of abstract non-empirical objects whose properties are open to non-empirical investigation, it may seem plausible to suppose that some other system of non-empirical abstract objects, time, is similarly open to *a priori* investigation.

In developing this suggestion, it will be enlightening to consider a similar thesis that has been thought to hold of sizes. For sizes, in virtue of being standardly identified by reference to physical objects (i.e., the size of the table), are like temporal items which are standardly identified by reference to things in time. The point that will be established on the basis of the most simplistic account of size would not be affected by considering a richer and more adequate account of size Thus, for the sake of simplicity, we will consider only the application of the notion of size to two-dimensional plane figures of some given shape. We apply to such figures the equivalence relation of being the same size as and we apply to them the comparative relation of being larger than. Our chief criterion for the application of these relations is based on the fit of a pair of such objects when placed on top of one another. We think of the relation of being larger than as having the properties of irreflexivity, transitivity, asymmetry. We cannot discover that this relation has these properties merely from knowing that its chief criterion of application is based on the fitting test. For we can envisage finding that object *A* fitted entirely inside object *B* and that object *B* fitted entirely inside object *C* and that, none the less, object *A* could not be fitted inside object *C* when this was attempted without the mediation of object *B*. If this result appeared to obtain we would not necessarily conclude that the relation of being larger than was a non-transitive relation. We might either suppose that the objects were not remaining constant in size through time or we might suppose that observational mistakes were being made about the results of the fitting tests. This thought experiment shows two things. First, it is part of the sense we attach to the relation of being larger in size than, that that relation is transitive and asymmetric. Second, to grasp the sense of this relation one must grasp that the fitting test is the chief criterion for the application of the relation. However, one could grasp this latter fact without grasping that it is also part of the sense of the term for that relation that the relation has certain properties such as transitivity. If objects behaved regularly as in the above thought experiment, or behaved in other relatively bizarre ways, our concept of size might cease to admit of any useful application. The linguistic framework we have evolved for talking about size has a utility that depends on certain general empirical conditions obtaining. If these conditions failed to obtain that concept might fail to admit of any useful application.

I have suggested that transitivity is part of the content of our concept of being greater than. Arguably there are other propositions about that relation whose truth reflects not so much empirical facts about the world but reflects further aspects of our concept of size. For instance, one might claim the following propositions to have such a status. Given any size there is a larger size (hereafter this proposition will be cited as *A*). One can envisage a reductionist account of size according to which the size of an object would not be construed as a non-empirical abstract object but would be construed as an equivalence class of equally-sized objects. In this case proposition *A* would at best express a contingent truth about the world. If, on the contrary, sizes are construed as non-empirical abstract objects, proposition *A* would be taken to be true, and necessarily so, in virtue of the character of the system of sizes. In developing this view, one might argue that the existence of a particular size is established non-contingently by showing that we understand what it would be for there to be an object of the size in question. Thus, on this approach, a given size is shown to exist by showing that it makes sense to suppose something might have that size. Since, given some particular size, we can sensibly suppose that something might have a larger size, we can say, on this theory, that it is a conceptual truth about size that for any given size there is a larger size.

We have a linguistic framework for talking about objects and their sizes that allows for any infinite iteration of terms for larger and larger sizes. On the view under consideration, nothing more is needed to establish the existence of these sizes over and above establishing that we could, in principle, recognize any one of these sizes to be possessed by some object. It has not been my intention either to defend this view of sizes nor to give an adequate exposition of this view. I have only sought to draw attention to some features of a plausible view of sizes. For seeing the plausibility, if not the tenability, of this view will assist us in seeing the plausibility of a similar view of time.

The specification of the content of our concept of time requires reference to certain temporal relations. The sense of terms for these relations is given, in part, by specifying the basic criterion of their applicability and, in part, by specifying the properties of these relations. For instance, reference needs to be made to the notion of being temporally before for which the basic criterion of application relates to memory impressions and the properties of which include asymmetry, transitivity, connectedness. One might also claim that specifying the sense of the tense operators is also involved and that this in turn is to be specified in part by giving syntactical and semantical rules for the use of such operators. One might expect to be able to argue from the properties possessed by the relations in question and the rules governing the tense operators to the conclusion that time had a structure of a certain kind.

Anyone taking this line of argument will presumably claim that time has no beginning. For the line of argument in question inclines him to treat time as possessing its properties as a non-contingent matter. As it is a contingent matter whether or not there was a first event it would not be plausible to argue at a conceptual level that time had a beginning. He can regard the claim that for any time there is an earlier time as being on a par with the claim that for any size there is a larger size. That is, it might be maintained that to establish the existence of a certain time, it is sufficient to show that we could recognize some event as having occurred at that time. Let t_0 be the time of an event E_0. Our linguistic framework for talking about time provides us with devices for referring to times before t_0. To establish that such terms have reference we do not need to establish that there were events having those times as their times of occurrence. We only need establish that we understand what it would be to discover that some event occurred at those times.

It is not implausible to construe proposition A about sizes as we did above. I do not wish either to imply that I think the doctrine in question is tenable; nor to imply that it is plausible to construe the time system on this model of non-empirical abstract objects. My concern is to uncover the basis of the common feeling that time is open to *a priori* investigation. It seems to me not impossible that someone who takes phrases of the form 'the colour of —' or 'the shape of —' as referring to non-empirical abstract objects, might construe the phrase 'the date of —' as referring to such an object.

Temporal items are abstract objects and it may be tempting to construe them as non-empirical objects. It may be that the feeling many have had that time could have no beginning stems from the above line of reasoning. Just as one might expect to reason from the concept of size to the structure of the system of sizes, one might expect to reason from the concept of time to the structure of the time system. However, the arguments of the preceding chapters have been adequate to refute the claim that one can argue from the concept of time to a particular topology for the time system. For we have seen that there are a range of different topologies that we might have grounds for ascribing to time. It might be objected that my arguments do not have this force on the grounds that the descriptions of contexts in which I said we would be warranted in asserting time to have a certain non-standard topology are descriptions of contexts which, if they obtained, would require a revision of our concept of time. That is, one might object to my line of argument by claiming that our *ordinary* concept of time is the concept of a time system with, say, the standard topology. There are two difficulties in establishing what the ordinary conception of time involves. For instance, we saw that we can give different representations of the rules covering tense operators. The different representations fit equally our ordinary linguistic practices and diverge on the structure implicitly ascribed to time. Secondly, in the absence of a precise criterion for distinguishing between conceptual revision and change in beliefs about matters of fact, it is empty to claim that, say, positing branched time instead of linear time represents a revision in the concept of time and not a change in our beliefs about the character of the time system.

6 THE FLEXIBLE RESPONSE THEORY

It has been argued that time need not have the standard topology. That is, we can without inconsistency or conceptual absurdity suppose time to have a structure which is not isomorphic to an open interval of real numbers. All the same, it may be remarked, the possible worlds in which we would be warranted in thinking of time as having some non-standard topology are extremely fanciful. And in view of that it might well be felt that the standard topology has some privileged position. It is not just one possible topology on a par with other possible topologies. Surely, it may be urged, there is some way of displaying the conceptually privileged character of the standard topology. While this feeling is, I think, common, I know of no articulated argument which would show that the feeling derives not merely from our habit of thinking in terms of the standard topology but from some feature of our conceptual system. (Of course, in claiming this I am setting aside those arguments which purport to show that time *must* have the standard topology.) I think that an argument worthy of consideration can be developed to support this feeling. In the end I will find the argument unsatisfactory. However, the argument to be given does reveal something of great interest about time.

The general sense of the argument can be conveyed most easily through a heuristic device. To that end, let us imagine that some omniscient and omnipotent Being has created a range of possible worlds. For each possible topological structure for time there is a world in the range which is such that one would be warranted in that world in asserting that time had that structure. For instance, there is a world in which one would be warranted in holding that time was open, a world in which one would be warranted in holding that time is closed, and so on. Let us imagine that the Being in question is to place you in one of those worlds. Without knowing which world you are to be placed in, you are to select one of the possible topologies for time. Once in the world you are required to stick by the topology you have selected. That is, your attempts to describe and explain the happenings in this world must take place within the framework ascribing to time the topology you selected in ignorance of a knowledge of the particular features of the world to which you are subsequently consigned. Which topology would it be best to select in this condition of ignorance? For reasons to be given, the best choice is the standard topology. If you have to decide, *a priori* so to speak, which topology to bring to bear on a world of unknown constitution, the rational thing to do is to select a topology that treats time as linear, unified, non-ending, non-beginning, dense or continuous. That this choice would in this special context be best can be seen by considering each aspect of the topologies for time in turn. Suppose, first, that one selects a topology which gives time a beginning. One will run into difficulties if one is placed in a world in which one would be warranted in believing in laws that admit of indefinite backwards extrapolation from the present state of that universe to indefinitely earlier states. If, on the other hand, one has selected a hypothesis of non-beginning time there will be no problem in that world. Nor will there be any particular problem in a world in which one would be warranted in asserting the existence of a first event. For in that case one will merely be lumbered with excess baggage, i.e., the range of times before the first event. Analogous considerations apply in the case of the choice between ending and non-ending time.

What about the choice between open and closed time? Suppose one opts for closed time and then is placed in a world which admits of continual novelty. In such a world, the laws admit of indefinite extrapolation from any given state to future and past states, no one of which is qualitatively identical to the given state, and it would be uncomfortable to be stuck with a closed-time framework. Suppose, instead, one were placed in the sort of possible world that was introduced in chapter III during the discussion on closed time. One who opted for closed time would be ideally placed in such a situation. However, as we saw, one who had opted for open time would be able to cope without difficulty. Under that topology, the world will be described as one in which history is precisely cyclical. Thus, one who opts for open time can deal with any possible world which can be dealt with by one who opts for closed time and in addition one who has selected open time can deal with worlds which cannot be dealt with adequately on the assumption of closed time.

If my argument of chapter VI concerning the choice between dense and continuous time is sound, in the sort of situation we are imagining there is no reason to favour continuity over denseness. For any world that can be dealt with adequately under the assumption of dense time can be handled with equal adequacy under the assumption of continuous time and vice versa. And this seems appropriate. For some of those who have felt that a particular topology has a privileged role have insisted only that time is at least dense: others have insisted that it is both dense and continuous.

Finally, consider the sort of world in which it was argued one would be warranted in supposing both space and time to be non-unified. One who sticks by unified linear time in such a case may be obliged to adopt particularly *ad hoc*, implausible and complicated hypotheses in describing and explaining the goings-on in such a world. This is certainly so if this possible world is dealt with by assuming that the claims about life in Harshland are mere delusions. However, the defender of the standard topology might argue as follows. Sticking by the hypothesis of unified linear time only means having to admit the possibility of time-travel and having to regard the temporal relations between Harshland and Pleasantville as unknowable in principle. Consequently, that world can be dealt with under the assumption of the standard topology. The defender of the standard topology can concede that the unknowability of the temporal relations he must suppose to obtain between events in Harshland and Pleasantville is an embarrassment. However, he can maintain that that embarrassment is much less extreme than the embarrassment of one who is committed to multiple time-streams in a world such as we believe the actual world to be. For such a person will be committed to the existence of a time-stream about which absolutely nothing can be known — it cannot even be known what, if anything, is happening in that time-stream. Thus, while the case for the standard topology with regard to the choice between unified and non-unified time may be less convincing than it is in regard to the other choices concerning the topology of time, there is certainly room for the advocate of this response to make a case. For while the standard topology is best for non-deviant worlds and while the non-standard topology is best for the deviant world, the standard topology is less unsatisfactory for deviant worlds than the non-standard topology is for non-deviant worlds.

This programme, which I will call the *flexible response programme*, claims for the standard topology a special position on the grounds that that topology can be deployed to deal more adequately with the entire range of types of possible worlds than can any other topology. Two factors can be cited in explanation of this interesting fact. First, the standard topology is richer than *many* rival topologies, in the sense that these can be *embedded* in the standard topology but not vice versa. For instance, consider a discrete, linear structure with first and last elements. Such a structure can be embedded in a linear continuous structure without first and last elements in the sense that there is a one–one order-preserving map from the former onto a proper sub-set of the latter. The converse does not hold. In addition, any linear dense structure can be embedded in any linear continuous structure. Also, any pair of non-intersecting linear structures (i.e., multiple time-streams) of some particular order type can be embedded in a single linear structure of that or a richer order type. Of course not all possible topologies for time are such that they can be embedded in this technical sense in the standard topology. A closed time topology cannot be so embedded. It is, however, the only possible topology I have considered that cannot be so embedded. The privileged position of the standard topology *vis-à-vis* closed time can be explained by reference to the second relative factor − the underdetermination of theory by data. For any context in which one would be warranted in holding that time was closed is a context in which one would be warranted equally in holding that time was open and that history was precisely cyclical. These two hypotheses in that context are underdetermined by all actual and possible observation. Thus, whenever there is a topological structure for time that cannot be embedded in the standard topology, the choice of that structure over the standard structure constitutes a case of the underdetermination of theory by data. These two factors, the relative richness of the standard topology and the underdetermination of data with regard to at least some features of time's topology, explain why it is that the standard topology has the privileged role it does have.

We rejected the view that one could establish *a priori* the topological structure of time. Interestingly, we have now seen that one can establish *a priori* the greater flexibility of the standard topology in dealing with all types of possible worlds. Thus, something survives of the view that the standard topology has a privileged position and that there is a role for *a priori* investigation of the topology of time. One can concede that this is both true and interesting without agreeing that we should treat time as having the standard topology regardless of the outcome of any possible investigations of the physical world. For the sense in which the standard topology has a privileged role is a very special sense. It is privileged only in the sense that it would be the best bet if we had to make a choice in the absence of any knowledge of the constitution of the world we are to deal with and, further, had to stick by that choice regardless of the outcome of our investigations of that world. It is just the flexibility of the standard topology that reveals the weakness of this move. For rather than adopt a framework that is more or less adequate to any possible world, we want to adopt the framework that is best for the specific world we find ourselves in. Perhaps one should start with

229

the assumption that time has the standard topology, but that assumption should be modified as necessary in the face of empirical investigation of the world in favour of the topological framework which is most adequate for dealing with the actual world.

7 THE FACTS OF THE MATTER

It has been argued that theories about the topology of time should be regarded as *empirical theories*. With some notable exceptions (among whom are Kuhn and Feyerabend), it is the fashion to espouse some form or other of a realist construal of empirical theories. In this section I consider whether theories about the topology of time should be given a realist construal. Those who style themselves as realists not uncommonly see their position as involving *at least* the following four ingredients. First, scientific theories are either true or false, and which a given theory is, it is in virtue of how the world is. In this context, a theory is to be thought of as the deductive closure of a set of postulates, and to speak of the truth or falsity of a theory is to talk of the truth or falsity of the conjunction of the postulates. This will be called the *ontological ingredient* in realism. A second and arguably not independent ingredient, which will be called the *causal ingredient*, is the claim that if a theory is true, the theoretical terms of the theory denote theoretical entities which are causally responsible for the observable phenomenon the occurrence of which is evidence for the theory. It would be of small comfort to learn that our theories are, indeed, either true or false representations of the world if we were precluded from being able to have rational grounds for believing that one theory is more likely to have one truth-value than the other. And our quite insatiable desire to know manifests itself in the third ingredient which will be cited as the *epistemological ingredient*. This is the claim that we can have warranted beliefs (at least in principle) concerning the truth-values of our theories.

Implicit in the characterization so far developed is the assumption that the goal of the scientific enterprise is the discovery of true explanatory theories. The reasonableness of this goal has seemed to some to be called into question by the fact that the history of science is a graveyard of falsified theories. Indeed, there seems to be evidence to support the meta-induction that any theory is found to be, strictly speaking, not true within, say, two hundred years of its being produced. Unless one is willing to take the courageous line of arguing that things are now, or will be, much different, the evidence points to the same fate for our current theories. And, thus, it looks as though the exercise of the epistemological power which the realist assumes we possess will always end in a negative verdict. This might incline us to wonder how rational it is to pursue a goal when the evidence is that that goal will never be realized. The realist counter-move is to argue that while all theories are false, some are falser than others. That is, the historically generated sequence of theories of a mature science may well be a sequence of false theories but it is a sequence in which succeeding theories have greater truth-content and less falsity-content than their predecessors. This empirical thesis, which will be called the *thesis of convergence*, would, if tenable, render it rational to pursue the goal of truth, for we would at least have some assurance that we were getting nearer the unobtainable goal. The thesis of convergence together with the ontological, causal and epistemological ingredients constitutes a sort of minimal common factor among the wide range of philosophers who in recent years have advocated a realist construal of scientific theories. These would include Boyd, Harré, Hesse and Putnam, among others.[10] Boyd, commenting on a paper of van Fraassen's,[11] has offered an explicit characterization of realism similar to that offered above:

> What realists really should maintain is that evidence for a scientific theory is evidence that both its theoretical claims and its empirical claims are 'on to something' about the way the world is. Being 'on to something' no doubt entails that a theory is in a certain respect importantly true, but it does not preclude its being in other respects profoundly wrong as well. Clearly, what the realist requires is a notion of approximate truth.[12]

We have seen in the course of our discussion several ways in which theories concerning the topology and metric of time may constitute cases of the underdetermination of theory by data. That is, with regard to some aspects of the topology and metric of time there may be incompatible theories which are empirically equivalent in the sense that no actual or possible observation could provide us with evidence favouring one theory over the other. Given that underdetermination obtains, realism as characterized cannot be accepted as the correct construal of the hypotheses concerning time that give rise to underdetermination. For given that all actual and possible data falsifies all theories for some aspect of the metric or topology of time except the empirical equivalent incompatible theories T_1 and T_2, the ontological ingredient in the realist position leads the realist to hold that there is something in virtue of which either T_1 is true or T_2 is true. However, *ex hypothesi*, nothing is going to count in favour of T_1 over T_2 and vice versa. In this context, nothing could count as evidence for thinking that T_1 is more likely to be true than false. This is incompatible with the epistemological ingredient in the realist position which holds out the hope (at least in principle) of having warranted beliefs concerning the truth-value of our theories. To use one of our examples — the realist wants to hold both that the world is such that either time is closed or it is not, and that we can come to have evidence concerning which it is. However, we have seen that in some contexts this is not possible.[13]

At this juncture something has to give. One might try, on the one hand, to weaken the ontological ingredient. Or, on the other hand, one might try to weaken the epistemological ingredient. Initially any one with realist sympathies inclines to respond to the dilemma by weakening the epistemological ingredient. One so inclined will insist that all scientific propositions have a determinate truth-value but, while maintaining that in most cases we can (in principle) have warranted beliefs concerning the truth-values of our propositions, it will be conceded that this does not hold for the non-empty class of empirically undecidable propositions. The reason this response seems to be the most plausible, indeed the *only* plausible response, lies in the following two factors. First, if one has any sympathy with a realist position one will have adopted a 'correspondence' theory of truth. Secondly, we tend to believe in the Law of Bivalence which amounts to the claim that any proposition is either true or false. We cannot abandon a 'correspondence' theory of truth without entirely extinguishing the spirit of realism. And given that there are at most two truth-values and that if a proposition has one truth value its negation has the other truth-value, a commitment to bivalence amounts to a commitment to the Law of the Excluded Middle (hereafter cited as LEM). Having robust common sense, we do not see how we can abandon LEM. For it is, after all, one of the *immutables* in virtue of being a law of logic. (If you can't believe that, what can you believe?) But these two ingredients basically constitute the ontological ingredient of realism. For example, by appeal to LEM we assert that either time is closed or it is not closed. And by appeal to the correspondence theory of truth, we conclude that there is something about the world in virtue of which one or other of these alternatives is true. So the only way out of the dilemma seems to involve supposing that there are *facts* for which we can have no evidence. Now this response to which we seem driven is not entirely implausible. For, surely, it might be retorted that it was a piece of not inconsiderable arrogance in the first place on the part of the human intellect to assume that all there is to be known can be known by finite beings such as ourselves. This response, the *Ignorance response*, involves maintaining that those propositions responsible for underdetermination are either true or false. It is conceded that with regard to these propositions we could not possibly have evidence concerning their truth-value. As such, this response involves embracing the possibility of *inaccessible facts* — facts for which we could have no information as to whether they do or do not obtain.

Alternatively, a realist might respond to the underdetermination of theories by restricting the scope of his realism in the following sense. Given a context in which some proposition P is empirically undecidable, the assumption that either P is true or P is false is withdrawn. As the realist holds that to be true (false) is to be true (false) in virtue of how the world is, this response involves dropping the assumption that there is something about the world in virtue of which P is true or something about the world in virtue of which P is false. That is, instead of supposing that there are inaccessible facts in virtue of which P is either true or false, we conclude that the world is simply indeterminate with respect to P. This response, the *Arrogance response*, amounts to holding that if we cannot know about something there is nothing to know about. One who makes the Arrogance response is embracing what was called in chapter VII the thesis of the essential accessibility of facts or *TEAC*.

Consider the example of an empirically undecidable proposition given in chapter III (the choice between closed and open time) in light of these alternative responses. Many have a strong inclination to say that in that possible world time either is closed or it is open — and that is that. Either the future occurrence of some state is a new and different occurrence of that state (i.e., time is open) or it is numerically the same occurrence (i.e., time is closed). Either time is such that it is like a closed curve or it is such that it is like an open curve. In making this response (the Ignorance response) we are taking underdetermination as pointing only to our inability to have evidence concerning which of these possibilities actually obtains. That is, we have a case of inaccessible facts. One who makes the Arrogance response will regard the heuristic device I introduced in the discussion of the possible world in chapter III, section 3, as inadmissible. The device in question involved us in imagining that we had been placed in one of a pair of possible worlds (in one of which time is closed and in one of which time is open). Such a world would be judged illegitimate by one inclined to make the Arrogance response on the grounds that one is just not entitled to make these stipulations. For there is no determinate state of affairs which either obtains or does not obtain which would make it true that time is closed. The set of facts constitutive of the one world, it would be claimed, *is the same as* the set of facts constitutive of the other world, and in that set of facts there is no fact answering to the proposition that time is closed and there is no fact answering to the proposition that time is open.

One who makes the Arrogance response in the face of an empirically undecidable proposition *P* will not be willing to assert that it is either the case that *P* or it is not the case that *P*, and is thereby committed to denying LEM to have the status of a genuine law of logic. On the other hand, one who makes the Ignorance response is likely to invoke the claim that LEM is a genuine truth of logic in attempting to justify his claim that there is a matter of fact at stake with regard to *P* — a matter of fact which is inaccessible. Consequently one inclined to the Arrogance response may well wish to avail himself of the interesting arguments Dummett[14] has explored for not asserting LEM. The consequences of this line of argument strike some as implausible. For there are many cases in which our intuitive inclination is to assert a substitution instance of LEM (i.e., either a city will be built at the North Pole some day or a city will never be built at the North Pole) where, given the line of argument explored by Dummett, we would not be entitled to do this. However, there is a weaker strategy which might be deployed by someone to vindicate the Arrogance response without embracing these apparently implausible consequences. To develop this strategy we need first to remember that we are dealing with propositions and contexts in which those propositions are empirically undecided in the sense that fixing the truth-value of *all* observation sentences leaves the truth-value of those propositions open. For instance, the first example of underdetermination provides such a proposition (that time is closed and history is unique) and such a context. One inclined, as I am, to the Arrogance response sees no reason to admit in such a context the requisite inaccessible facts. Thus, without asserting that there can be no inaccessible facts, one asks of one making the Ignorance response what his grounds are for asserting that there are inaccessible facts which either make it true that time is closed, etc., or make it true that time is open, etc. (we are assuming that all other alternatives have been excluded so

234

that the only way in which time can fail to be closed is if it is open).

It will not do for the advocate of the Ignorance response to appeal to LEM. For what is in question is just whether LEM holds for empirically undecidable propositions. Consequently, it is not clear what could possibly count as a reason for thinking that there is a matter of fact (a matter of inaccessible fact) at stake here. For there is nothing that would be explained by the supposition that there are such facts. And that being so, we should prefer the ontologically weaker position which does not assert the existence of such facts. This means restricting the scope of LEM to exclude empirically undecidable propositions. If the underdetermination of theory by data is a relatively rare phenomenon this will not mean a very extensive restriction. The limited scope of the restriction arises from the assumption that there is a matter of fact at stake with regard to any observational proposition. Given this assumption we can still assert, for example, that either a city will be built some day at the North Pole or a city will never be built at the North Pole. For a distribution of truth-values over the set of all observational propositions will determine the truth-value of 'There will be a city built some day at the North Pole'. One who denies this assumption will have to embrace a more extensive restriction on LEM.

To opt for the Arrogance response means ceasing to regard empirically undecidable propositions as expressing hypotheses about the facts. Consequently, the onus is on the proponent of the Arrogance response to give an account of the role those propositions play in the theories that contain them. One possibility is that these propositions should be seen as serving to specify a mode of description or general framework within which the hypotheses about the facts are to be expressed. That is, for instance, one who asserts that time is continuous is not making a guess about the facts but is opting for a particular net for catching the facts. All the facts that there are can be expressed in terms of this framework; or, equally, they can be expressed in terms of the rival framework based on the treatment of time as merely dense. This possibility will have to be explored elsewhere. My concern now is only to note the need of some such account.

8 TIME AS A THEORETICAL STRUCTURE

In this section I shall outline a general picture of time that is compatible with the results of the discussion to date and which would be appropriate if one opts for the Ignorance response. This general picture of time as a theoretical system or structure of quasi-abstract temporal items is modelled on a version of the realist construal of the role of theoretical terms in scientific theories. Thus, it will be fruitful to begin with a brief consideration of the realist construal of theoretical entities. The position which I will sketch briefly is controversial and within the confines of this work I cannot defend the view. In any event, I will not be arguing from this theory to a picture of time. For this theory is used simply to suggest a view of time that could be defended independently of the general view of theoretical terms.

In what follows I will assume, as in chapter III, that one can at least distinguish degrees of theoricity. When reference is made to theoretical (observational) terms, that is to be understood as referring to terms towards the theoretical (observational) end of the spectrum of terms used in science. In developing a characterization of a realist position with regard to theoretical terms in science, let us think in terms of standard scientific discourse about sub-atomic particles. For the realist, our interest in having a theory of sub-atomic particles stems from a desire to provide an explanation of certain observable facts. For instance, Thompson posited the evidence of electrons and hypothesized about their properties in order to explain scintillations produced in a cathode-ray tube. Unlike the instrumentalist who seeks only to predict observations, the realist wishes to explain observations and in so aiming seeks theories which are true. For a theory can only provide the basis of a good explanation of a phenomenon if the theory is true or approximately true. While realists differ in the accounts they give of a correct theory for the meaning of theoretical terms, they are united in holding that evidence for the empirical adequacy of a theory is evidence for its truth or approximation to the truth. That is, successful prediction is not an end in itself but is the sign of truth or approximation to the truth. Evidence that one theory is more empirically adequate than its rivals provides, according to the realist, grounds for holding the theory to be true or to be the best available approximation to the truth. For the realist, accepting the theory as tentatively true commits him to the existence of whatever has to exist in order for the theory to be true. Thus, for example, adopting a theory of sub-atomic physics commits us to the existence of electrons, etc., whether or not these entities are, or ever shall be, observed.

In applying the realist model to time I will use the notion of a *theory of time* in a somewhat different sense than hitherto used. Providing a theory of time for some possible world in this sense will involve ascribing to time a particular topological and metrical structure *and* specifying the relation between the history system of that world and the time system. We can locate within such a theory the analogue of the theoretical postulates of a theory which link theoretical terms. For instance, these might include the assertion that space and time are of the same order type. There will also be analogues of the mixed bridging postulates that link theoretical terms with observational terms. For instance, these might include the claim that if time is closed, for any given state of the universe there will be in the future a qualitatively identical state.

If we take it in the context of some possible world that some theory of this type is literally true we will be committed to the existence in that possible world of a system of temporal items (these may be instants, say) having the character represented in the theory. The system of items whose existence is presupposed will not be concrete objects in the sense in which the sub-atomic particles are concrete objects (i.e., that they are occupants of space and time which are in principle open to instrument-assisted observation). The items in question will be best described as quasi-abstract for the reasons given in chapter VI. By allowing that there can be theories of this kind we are rejecting a classical dichotomy between abstract objects, the properties of which are investigated *a priori* (such as mathematical objects), and concrete objects, the properties of which are investigated *a posteriori* (such as sub-atomic particles). For the investigation of the time system is an empirical matter. However, the system of items constituting time are, in virtue of so constituting time, distinct from the types of item that can occupy space and time.

We noted in the context of our brief account of sub-atomic theories that we are interested in having such theories in order to explain observable phenomena. And a precondition of such a theory being genuinely explanatory is its being true or approximately true. Similarly we are interested in having a theory of time for explanatory reasons, and in regarding it as genuinely explanatory we are committed to regarding it as being true or approximately true. In this case what we wish to describe and explain are the observable characteristics of the history system of the possible world we are dealing with. The ways in which a particular theory of time will be explanatory depend on the particular world and theory in question. For instance, in the sort of world considered in the discussion of non-unified time and non-unified space, the theory that time is non-unified will explain in part why it is that the inhabitants of Pleasantville tell consistent and mutually supporting stories of life in Harshland. Or, to take another example, a theory of time for the world of chapter II will suppose that time has the standard topology but is richer than the history system. Such a theory will provide part of the explanation of the systematic character of the system of vanishing functions governing that world. In the case of a scientific theory we seek the weakest theory which is both compatible with all observation and is yet genuinely explanatory. Similarly in the case of time we seek the weakest theory of time which fits with the observable character of the history system while allowing the best explanation of that system. For that reason, if we thought there was a first event we would not adopt a theory of time that involved positing the existence of times before the first event.

It is interesting to note that there could be possible worlds in which there would be no reason to posit the existence of a time system at all. For instance, consider a possible world in which nothing warrants asserting the existence of temporal vacua and in which Russell's postulates on the set of events are satisfied (cf. p. 128). In that case we might talk of temporal items and the time system. However, we could introduce the notions of both temporal items and time system in a purely reductionist manner. There is no explanatory aim that would be served by positing the existence of temporal items and a time system over and above the system of equivalence classes of events. This is the result we would expect if my analogy between theories of time and theories of sub-atomic particles is appropriate. For there are possible worlds in which there would be no grounds for positing the existence of sub-atomic particles. Similarly, there are possible worlds in which there would be no grounds for positing the existence of a system of quasi-abstract temporal items over and above the system of events.

Attractive as this picture of time as a theoretical system of quasi-abstract items may be, it has its unhappy aspects. For, given my arguments concerning the underdetermination of theory by data, there will be possible worlds for which there are rival theories of time and there will be no observation or experiment which will favour one theory over the other. Which theory is true of that world is a matter on which we could have no warranted beliefs. This makes time simply intractable and thus mysterious. But perhaps that is the way it is. Undoubtedly we have powerful intuitions inclining us to think that the facts constituting the world either, say, make it true that time is dense, or the facts make it true that time is continuous. And until we have a proof drawn from a viable general theory of language that such intuitions are ill-founded we cannot simply dismiss this general picture of time. Time *is* somewhat mysterious and those impressed by this may feel that it should not surprise us that a certain element of mystery intrudes here. Indeed, they may feel that one should distrust any theory of a major aspect of time that does not find room for elements of the mysterious and the ineffable.

9 TIME AS A THEORETICAL FRAMEWORK

A reason, which is admittedly not compelling, has been given for favouring the Arrogance response to underdetermination. The sort of construal one who makes this response would give of hypotheses concerning the metric and topology of time which give rise to underdetermination can be extended to all hypotheses concerning the topology and metric of time to generate what will be called the picture of time as a *theoretical framework*. On this picture, hypotheses about time are not seen as representing hypotheses about the facts but are seen as specifying modes of descriptions for dealing with the facts. For instance, one might construe the assertion that time is continuous as tantamount to a methodological rule or regulative principle recommending that we adopt a certain kind of framework for describing the facts. All the facts that there are can be expressed in terms of this framework; or, equally, they can be expressed in terms of the rival framework based on the treatment of time as merely dense.

Adopting the view of time as a theoretical framework places constraints on the view one can take of ontological assertions relating to time. Let us suppose that one who adopts this view commits himself to, say, the assertion that time has some particular topological structure. He is thereby committed to the assertion that there is a system of temporal items having that structure. If the assertion that time has the structure in question lacks truth-conditions so will the assertion that there is such and such a system of temporal items. There is a general view of ontology, due to Carnap, which provides a construal of existential statements which will be particularly amenable to one who adopts the view of time under discussion.

According to Carnap, introducing devices into a language for talking about a type of item is adopting a linguistic framework. With regard to items of the kind in question one can ask internal and external ontological questions. Internal questions are answered through techniques specified when the linguistic framework is specified. For instance, Carnap considers the linguistic framework introduced in the context of introducing terms for natural numbers. The question as to whether or not there is a prime number greater than 100 is taken as an internal question. According to Carnap the rules specifying the linguistic framework in this case entail that there is a prime number greater than 100. To ask, on the other hand, whether or not there really are natural numbers is, for Carnap, to ask an external question whose only legitimate construal is as a question concerning the practical utility of adopting the linguistic framework in question.

Carnap applies this thesis to the spacetime points of physics:

> The step from the system of things (which does not contain spacetime points but only extended objects with spatial and temporal relations between them) to the physical coordinate system is again a matter of decision. Our choice of certain features, although itself not theoretical, is suggested by theoretical knowledge, either logical or factual. . . . Internal questions are here, in general, empirical questions to be answered by empirical investigations. On the other hand, the external questions of the reality of physical space and physical time are pseudo-questions. A question like 'Are there (really) spacetime points?' is ambiguous. It may be meant as an internal question; then the affirmative answer is, of course, analytic and trivial. Or it may be meant in the external sense. 'Shall we introduce such and such forms into our language?'; in this case it is not a theoretical but a practical question, a matter of decision rather than assertion, and hence the proposed formulation would be misleading. Or finally, it may be meant in the following sense: 'Are our experiences such that the use of the linguistic forms in question will be expedient and fruitful?' This is a theoretical question of a factual, empirical nature. But it concerns a matter of degree; therefore a formulation in the form 'real or not?' would be inadequate.[15]

A general discussion of the Carnapian approach to ontology cannot be undertaken within the confines of this work. I have sought only to show that one who opts for the view of time as a theoretical framework is committed to something like the Carnapian view of ontology, *at least as restricted to assertions about time*. I would, in fact, argue that the Carnapian view of ontology is not tenable with regard to any range of propositions that ought to be treated as being true or false in virtue of how the world is. However, if propositions (such as those relating to the topology and metric of time on the theory being considered) which are not to be so construed commit those who assert them to existential assertions, the Carnapian construal of such assertions is appropriate.

If one views time as a theoretical framework, one is rejecting the claim that there is some determinate independently existing time system the character of which makes hypotheses about it true or false. Assertions about the existence of temporal items can then only be construed in a Pickwickian way (either as 'internal assertions' or as 'external assertions'). Herein lies the implausibility of the view of time as a theoretical framework. For we saw in chapter II that there are contexts in which one would want to vest temporal items with real causal powers. But on the theory being considered, while things in time have real existence, temporal items and the time system do not have real existence. For assertions about time only provide a framework within which to organize assertions about things in time. Such assertions are not thought of as being true or false in virtue of some independently existing system of temporal items.

The difficulty with the theoretical framework theory is that it gives us too few facts. It does not allow that there can be a matter of fact at stake in two different sorts of contexts in which we would wish to hold that there is a matter of fact at stake. First, there are contexts which would include those where the assertions about the topology of time are decidable (i.e., in a world in which evidence supports the hypothesis that there is no first nor last event and that history is not cyclical, the evidence supports the claim that time is linear, non-ending and non-beginning). In such a context, we should construe the assertions realistically. Second, there may be contexts in which we wish to vest temporal items with causal powers. This requires according such items an ontological status that the theory precludes. While the theoretical framework gives us too few facts, the theoretical system theory gives too many facts. For we saw that such a theory is committed to inaccessible facts in the context of underdetermination and there is no reason to admit the occurrence of such inaccessible facts.

The way out of this impasse is to adopt the view of time as a theoretical system *up to underdetermination*. That is, for any assertions about the topology of time and any context in which those assertions are decidable we should hold that there is a matter of fact at stake. If in a given context all such assertions are decidable, there is an existing time system with a determinate structure in virtue of which those assertions are true or false. If some, but not all, such assertions are decidable there is a partially determinate time system in virtue of which the decidable assertions are true or false. The undecidable assertions in such a context are not treated as being true or false in virtue of the independently existing time system but as only providing a framework within which to express hypotheses about the facts concerning the history system. Further, in any context in which the evidence requires the supposition of temporal items with causal powers, we treat these temporal items as really existing.

This general strategy of adopting a realist construal up to under-determination is, I would argue, the appropriate strategy to adopt in the case of *all* physical theories. If this position can be sustained it means that, in a sense, the investigation of time is on a par with the investigation of the other subject matters which are the province of physics, say, the structure of the nucleus. For in the case of time and of the structure of the nucleus, the choice between theories is to be decided by considering what it is best to posit in order to explain the observations of things in time. In both cases, those theories or aspects of theories that are not underdetermined are to be construed realistically. None the less, time does differ from other subjects of physical investigation. On the one hand, it is much more indirectly connected with observable phenomena and this means that hypotheses about the structure of time have much less empirical content than, say, hypotheses about the structure of the nucleus. On the other hand, time is ultimately rather simple. There is no other subject matter for physics that admits of such phenomenally simple theories. Of course, freedom from complexity is not the same as freedom from mystery. And even if the views advanced in this work do make some aspects of time less mysterious, its promiscuous character means that there are depths yet to be plumbed; in particular, the perhaps most puzzling aspect of time, the relation between time and consciousness, remains.

APPENDIX

PROPERTIES OF RELATIONS

Reflexivity
A relation R defined on a set A is reflexive if and only if every object in A bears R to itself. If we restrict the range of the universal quantifiers to the set A we can express this condition as: $(x)(Rxx)$. In the following definitions we will make this restriction. The relation of being as tall as defined on the set of all living persons is reflexive.

Irreflexivity
R is irreflexive if and only if $(x)(-Rxx)$, e.g., the relation of being taller than on the set of living persons.

Symmetry
R is symmetrical if and only if $(x)(y)(Rxy \rightarrow Ryx)$, e.g., the relation of being as tall as on the set of living persons.

Asymmetry
R is asymmetrical if and only if $(x)(y)(Rxy \rightarrow -Ryx)$, e.g., the relation of being taller than on the set of living persons.

Anti-symmetry
R is anti-symmetrical if and only if $(x)(y)(Rxy \ \& \ Ryx \rightarrow x = y)$, e.g., the relation of being less than or equal to on the set of integers.

Transitivity
R is transitive if and only if $(x)(y)(x)(Rxy \ \& \ Ryz \rightarrow Rxz)$, e.g., the relation of being as tall as on the set of living persons.

Connectivity

R is connected if and only if $(x)(y)(x \neq y \to Rxy \lor Ryx)$, e.g., the relation of being less than on the set of integers.

Equivalence Relations

R is an equivalence relation if and only if R is reflexive, symmetrical and transitive, e.g., being as tall as.

Equivalence Classes

If R is an equivalence relation of a set A, and a is in A, then the equivalence class of a is the set of members of A all of which bear R to a.

Partition

A partition of a set A is a set of sub-sets of A such that any member of A is in at least one of the sub-sets and no element of A is in more than one of these sub-sets. It is easily shown that any equivalence relation defined on a set generates a partition of that set (the set of all equivalence classes) and that any partition generates on equivalence relation (being in the same sub-set as).

Ordering

R is an ordering relation if and only if $(\exists x)(\exists y)(x \neq y \ \& \ Rxy \ \& \ -Ryx)$, e.g., the relation of being heavier than on the set of elephants.

Partial Ordering

R is a partial ordering if and only if R is reflexive, anti-symmetrical and transitive, e.g., being an integral multiple of on the set of positive integers.

Linear Ordering

R is a linear ordering if and only if R is a connected partial ordering, e.g., being equal to or less than on the integers.

Discreteness

An ordering relation R is discrete if and only if $(x)(y)[Rxy \ \& \ x \neq y \to (\exists z)(Rxz \ \& \ x \neq z \ \& - (\exists w)(Rxw \ \& \ Rwz))]$. If an object bears R to something then there is a unique next object to which the object bears R, e.g., the relation of being less than on the set of integers.

Density

R is a dense ordering if and only if $(x)(y)(Rxy \to (\exists z)(z \neq x \ \& \ z \neq y \ \& \ Rxz \ \& \ Rzy)$, e.g., the relation of being less than on the rationals.

Upper Bound

If R is a partial order on A and B a sub-set of A, then an object a in A, $a \epsilon A$, is an upper bound for B if and only if $(x)(x \epsilon B \to Rxa)$.

Appendix

Lower Bound
If R is a partial order on A and B is a sub-set of A, then $a \epsilon A$ is a lower bound for B if and only if $(x)(x \epsilon B \rightarrow Rax)$.

Least Upper Bound
If R is a partial order on A and B a sub-set of A then a is a least upper bound for B if and only if a is an upper bound for B and a bears R to all upper bounds for B.

Greatest Lower Bound
If R is a partial order on A and B a sub-set of A then a is a greatest lower bound for B if and only if a is a lower bound for B and all lower bounds bear R to a.

Cut
If R is a partial ordering on A and B a sub-set of R then B and its complement, $\bar{B}$ (i.e. the set of all elements of A not in B) are a cut of A if and only if either no element in B bears R to any element in $\bar{B}$ and some element in $\bar{B}$ bears R to some element in B or no element in $\bar{B}$ bears R to any element in B and some element in B bears R to some element in $\bar{B}$.

Continuity
R is a continuous ordering on A if and only if R is a partial ordering and any cut B, $\bar{B}$ of A where some element of B bears R to some element of $\bar{B}$ is such that either B has a least upper bound in $\bar{B}$ or B has a greatest lower bound in $\bar{B}$, e.g., the relation of less than or equal to on the set of all real numbers.

Bijection
A function f from a set A onto a set B is a bijection if and only if for each element a in A there is one and only one element b in B such that $f(a) = b$ and for any element b in B there is an element of a in A such that $f(a) = b$.

Isomorphic
A set A ordered by a relation R is isomorphic to a set A' ordered by a relation R' if and only if there is a bijection f from A to A' such that $Rab \rightarrow R'f(a)f(b)$.

NOTES

I THE NATURE OF TIME

1 Augustine, 1948, Book XI, sect. x.
2 Waismann, 1959, p. 117.
3 See in this regard Bouwsma, 1967, 'The Mystery of Time'.
4 Leibniz, 1956.
5 More recently the reductionist view of time has been defended by the following, among others: Bunge, 1967; Hinckfuss, 1975; Russell, 1914.
6 Arguably the reductionist needs a more complicated construction. This will be considered later in ch. VI (sect. 5) and ch. VIII (sect. 5).
7 Quoted from the translation given in Čapek, 1975, pp. 203–4.
8 ibid., p. 208.
9 Swinburne, 1968.
10 ibid., p. 209.
11 In ch. III, sect. 5, an account will be given of the conditions under which a reductionist would have to hold time to be closed rather than open.
12 For an introduction to these issues that I will not be exploring one can consult the works of Dummett, Gale and Prior which are cited in the bibliography.

II TIME AND CHANGE

1 For the moment I am assuming the framework of classical physics. Within the framework of the Special Theory of Relativity a considerably more complicated construction of temporal items is required. See in this regard sect. 5 of ch. VIII.
2 'For when the state of our own minds does not change at all, or we

have not noticed it changing, we do not realize that time has elapsed', Aristotle, 1928, 218^{b}21-4.

3 The term 'temporal indexicals' covers both tenses and temporal demonstratives such as 'now'. See in this regard Reichenbach, 1947.

4 Blackburn, 1973, ch. 4.

5 Shoemaker, 1969, pp. 363-79.

6 Van Fraassen, 1970, p. 28.

7 Shoemaker, 1969.

8 While I independently developed my own argument as an extension of one I had used in connection with a consideration of the intelligibility of the notion of absolute space, some features of the presentation of my argument have been influenced by my subsequent reading of Shoemaker's important article.

9 Shoemaker, 1969, p. 378.

10 See in this regard Hill, 1955. In part, Hill is concerned to tentatively explore the treatment of changes in energy and momentum as discrete against a background of continuous spacetime. See also Bohm *et al.*, 1970. The authors are here concerned to introduce a framework in terms of which physical quantities could be treated as changing discretely.

11 Poincaré, 1913, p. 188.

12 Harré, 1970.

13 Boscovitch, 1966, p. 23.

14 For a more detailed exposition of the notion of continuous time see sect. 2 of ch. VI.

15 Harré, 1970, p. 289.

16 That this is Harré's thought is further illustrated when, in discussing the thesis that space and time must be of the same order type, he states: 'But since it is a real particle it must be demonstratively at s1 or s2, actually and it cannot be arbitrarily set at either' (Harré, 1970, p. 290).

17 For an explication of these terms see the Appendix, p. 245.

18 This is a controversial claim. However my substantial point can be made if one prefers to employ any of the standard rival models of explanation.

19 Maxwell, 1953, p. 13.

20 Frege (1971, p. 203) formulates the maxim as follows: when identical circumstances arise, we expect the same consequences to occur irrespective of when they happen. If other consequences occur, we infer that we failed to notice all the relevant circumstances; we do not put the blame on the time difference as such.

21 Lucas, 1973, p. 72.

22 This conclusion may be too strong. See p. 35.

23 Quine, 1960, pp. 258-9.

24 See in this regard Benacerraf, 1965.

25 Quine, 1969, p. 263.

26 Bennett has argued that Leibniz's grounds for denying the possibility of spatial vacua were moral and theological in character.

According to Bennett, Leibniz 'rightly did not think that the possibility of vacuum is ruled out by the relational theory of space'. In support of this contention Bennett cites Leibniz's remark in his Third Reply to Clarke (para. 5) that space 'is nothing at all without bodies, but the possibility of placing them'. Leibniz's considered view may have been that all assertions about space are equivalent in meaning to assertions about actual *and* possible spatial relations between actual *and* possible objects. Given that this is what is to be understood by the term 'relational theory of space', the proponent of a relational theory of space can consistently acknowledge the possibility of spatial vacua as will be shown in this section. See Bennett, 1974, sect. 48.

27 In considering Leibniz's theory of space, Bennett has suggested construing, for instance, the statement that there is an unoccupied location equidistant from objects *a, b,* and *c* as equivalent to the conjunction of the statement that there is no object equidistant from *a, b,* and *c* and the statement that it is causally possible for some object to be equidistant from *a, b,* and *c*. Bennett fails to note the difficulty in this analysis here being outlined. See Bennett, 1974, p. 147.

28 Hooker, 1971. Hooker refers to what I have called the reductionist theory as the Relationalist theory.

29 ibid., pp. 110–11.

III THE TOPOLOGY OF TIME I: THE LINEARITY OF TIME

1 For an introduction to mathematical topology see Mendelson, 1968.

2 For a definition of this and other properties of relations used here and elsewhere in this text see the Appendix, pp. 243–5.

3 I follow Prior (1967) in this use of the notion of proposition which diverges from the more standard notion of a proposition as something whose truth-value does not change with time.

4 For a useful survey of first-order axiom systems and tense-logical postulate sets of this sort see Appendix A of Prior, 1967.

5 Putnam, 1975a, 205. See also in this regard Prior, 1967, p. 59.

6 Swinburne, 1968, p. 209.

7 Prior, 1967, p. 75.

8 ibid., p. 64.

9 ibid., pp. 198–9.

10 A proof is given for symmetry and transitivity in Resher and Garson, 1971, p. 118. The result for reflexivity can be similarly established.

11 Rescher and Garson, 1971, p. 120.

12 Newton-Smith, 1978.

13 This conception of cyclical history, popular with the Stoics, is neatly expressed by Chrysippus as follows. 'It is not at all impossible

that after our death, after the passage of many periods of time, we shall be re-established in the precise form which we now have. . . . There will again be a Socrates, a Plato, and each man with the same friends and the same fellow citizens, and the restoration will not take place once but many times; or rather all things will be restored eternally' (Manuel, 1965, p. 9).

14 Quine, 1970, p. 179.
15 Putnam, 1975b, p. 180.
16 For a proof of this result see English, 1973.
17 Swinburne, 1968, p. 51.
18 Dummett, 1973, p. 617 fn.
19 Eudemus, 1870, p. 74.
20 Grünbaum, 1973a, pp. 197–8.
21 Dummett, 1973, p. 543.
22 Swinburne, 1968, p. 109.

IV THE TOPOLOGY OF TIME II: THE UNITY OF TIME

1 Quinton, 1962, pp. 130–47.
2 Swinburne, 1968, ch. 2, sect. 10.
3 See pp. 58 ff.
4 See in this regard Prior, 1967, p. 199, where it is argued that diagrams 2 and 3 but not 1 represent genuine possibilities for the reasons I noted on p. 59.
5 Swinburne, 1968, p. 200.
6 Hollis, 1967.
7 ibid., p. 530.
8 Icabod and Isabel communicate with one another when they are together in Pleasantville. Consequently, during a visit to Harshland, Icabod may have foreknowledge concerning his own future actions in Harshland. Some may argue that this sort of foreknowledge limits Icabod's freedom of action. I would argue that this is not so as has been effectively demonstrated in Lewis, 1976.
9 As the term 'memory' is standardly used, that I remember doing X at time t entails that X was done before time t. Consequently in according objective status to the accounts which Icabod and Isabel provide in Pleasantville of life in Harshland we cannot describe their state as a state of remembering. We will have to introduce a term that functions like memory save for the temporal implication in question.
10 Quinton, 1962, p. 146.
11 Hollis (1967) offers this objection.
12 Swinburne, 1965. Swinburne has subsequently argued that unified space and non-unified time is incoherent. See Swinburne, 1968, ch. 4, sect. 10.
13 This point was made to me in discussion by Dr R. C. S. Walker.

V THE TOPOLOGY OF TIME III: THE BEGINNING OF TIME

1 Swinburne, 1968, p. 207. Actually Swinburne offers the future analogue of this argument and claims the past analogue of his argument holds. In the quoted passage, I have transposed his argument to the past time case. Swinburne claims the argument given above establishes unboundedness. To establish that time is non-ending and non-beginning requires the additional premise that time is not closed. Swinburne argues that time of logical necessity is not closed. In what follows I assume we are dealing with non-closed time and hence treat his argument as an attempt to show that time of logical necessity is non-beginning.

2 Aristotle (1938), *Physics*, VIII, $251^b 19$–23.

3 In this regard see Findlay (1941) who opts for a stronger postulate than (i) that all events, past, present and future, will be past. He would presumably also opt for the postulate that all events (past, present and future) have been past, which is stronger than (ii).

4 Kant, 1963, p. 75.

5 Quinton, 1973, p. 88.

6 Wittgenstein, 1975, p. 309.

7 See in this respect Clarke's Fourth Reply in Leibniz (1956), and Wittgenstein, 1975, p. 163, where he claims that 'empty infinite time is only the possibility of facts which alone are the realities'.

8 Peirce, 1935, pp. 224–5.

9 Leibniz, 1956, p. 27.

10 See in this regard Harré, 1962.

11 Kant, 1963, p. 397.

12 This universe characterizes the Einstein–de Sitter model of the universe. See North, 1965, ch. 6, for an account of this and related cosmological models.

13 Indeed, there are physical theories according to which the transformations around the time of $t+$ determined the current properties of matter. On this account of the matter we should see time $t+$ as the limits of extrapolation on the basis of these laws. For before $t+$ the constituents of the universe and their behaviour are governed by other laws. In this regard see the work of C. Hayashi cited by Gamov, 1954, p. 63.

14 Leibniz, 1956, p. 75.

15 See in this regard Hawking and Ellis, 1973.

VI THE TOPOLOGY OF TIME IV: THE MICRO-ASPECTS

1 Nerlich, 1976, p. 178.

2 Whitrow, 1961, p. 156.

3 Nerlich, 1976, ch. 9.

4 Poincaré, 1946, p. 49.

5 Harré, 1970, pp. 289–90.

6   For an introduction to the precise axiomatization of Newtonian mechanics see Suppes, 1957, ch. 12.
7   Zahar, 1973.
8   Bunge, 1967, p. 95.
9   Russell, 1914, pp. 123-8; 1936, pp. 216-28.
10 Van Fraassen, 1970, p. 106.
11 Russell, 1914, lecture IV.
12 ibid., p. 126.
13 Prior, 1968, pp. 122-3.
14 For a detailed account of this reduction see Stoll, 1961b, pp. 137-42.
15 See ch. III, sect. 2.
16 Prior, 1968, pp. 109-12.
17 ibid., p. 112.
18 Tarski, 1956, pp. 24-9.
19 This means that we are using the notion of being a proper part of. Tarski's basic notion is that of being a part of or being equal to (Tarski, 1956, p. 25).
20 For an alternative construction of instants which are provably continuous see Hamblin, 1971.
21 Dummett, 1973, ch. 14.
22 See in this regard Wheeler, 1962, and Graves, 1971.

VII THE METRIC OF TIME

1   Reichenbach, 1950, ch. 2, sect. 16-18.
2   Reichenbach, 1950, pp. 116-17.
3   ibid., p. 116.
4   Grünbaum, 1973a, pp. 59-70.
5   Poincaré, 1946.
6   Reichenbach, 1950.
7   Grünbaum, 1968 and 1973a.
8   Putnam, 1975b, p. 165.
9   Grünbaum, 1973b, pp. 505-7.
10 For the standard proof that the set of rational numbers is denumerably infinite and that the set of real numbers is not, see Hunter, 1971, pp. 17-25.
11 Grünbaum, 1973b, p. 519.
12 Newton, 1953, p. 17.
13 Swinburne, 1968, p. 209.
14 Bruno, *Camaeracensis Acrotismus*, art. xxxviii-xl. Translated in Čapek, 1975, pp. 191-2.
15 Leibniz, 1956, p. 52.
16 ibid., p. 75.
17 Lucas, 1973, pp. 90-1.

VIII THE SPECIAL THEORY OF RELATIVITY

1 Zeeman, 1964.
2 For the contrary view see Ellis and Bowman, 1967. See also the critiques of Ellis and Bowman in Grünbaum *et al.*, 1969.
3 Swinburne, 1968, ch. 11.
4 Prior, 1967, pp. 203–7.
5 ibid., pp. 205.
6 I have developed a metrical tense logic (which gives time a linear structure) that is rich enough to contain the STR (Newton-Smith, 1979).
7 Quine, 1960, p. 172.
8 For an articulation of the feeling that there is a tension see Prior, 1970.
9 Dingle, 1964, p. 46.
10 North, 1970.
11 See in this regard Grünbaum, 1973a, p. 345.

IX THE DIRECTION OF TIME

1 See in this regard Gale, 1968, ch. VI.
2 Dummett, 1964. See also Mackie, 1974, ch. 7.
3 Reichenbach, 1950 and 1956. For a critique of Reichenbach's mark method and an exposition of his causal theory see Grünbaum, 1973a. Grünbaum's theory has in turn been criticized in Lacey, 1968.
4 See in this regard Gale, 1968, p. 131.
5 For an introduction to these issues see Grünbaum, 1973a; Gold, 1967; Popper, 1956, 1957 and 1958; Sklar, 1974.
6 Boltzmann, for example, described the distinction as 'a mere illusion arising from our specially restricted view point' (Boltzmann, 1964, p. 446).

X TOWARDS A POSITIVE THEORY

1 Van Fraassen, 1970.
2 Where Frege wishes to banish from the philosophy of mathematics both empirical contamination and psychological contamination we banish psychological contamination and insist on empirical contamination. See Frege, 1950.
3 Mach, 1960.
4 Wittgenstein, 1961.
5 Van Fraassen, 1970, p. 102.
6 ibid., p. 102.
7 Wittgenstein, 1961, 2.0251.
8 ibid., 6.3751.

9   Van Fraassen, 1970, p. 185.
10 See in this regard Hesse, 1974, p. 290; Harré, 1970, p. 90; Boyd, 1973; Putnam, 1975/6.
11 Van Fraassen, 1976.
12 Boyd, 1976, pp. 633–4.
13 This tension between underdetermination and realism has been noted by Glymour, 1967.
14 See Dummett, 1958–9; 1969; 1973; 1978.
15 Carnap, 1956, pp. 212–13.

BIBLIOGRAPY

Aristotle (1928), *The Works of Aristotle*, ed. W. D. Ross, vol. 8: *Physics* (Oxford: Clarendon Press).

Atkinson, D., and Halpern, M. B. (1967), 'High Energy Scattering', *Journal of Mathematical Physics*, 8, pp. 373–87.

Augustine (1948), *Confessions*, trans. E. B. Pusey (Chicago: Henry Regnery).

Benacerraf, P. (1965), 'What Numbers Could Not Be', *Philosophical Review*, 54.

Bennett, J. (1971), *Locke, Berkeley, Hume* (Oxford University Press).

Bennett, J. (1974), *Kant's Dialectic* (Cambridge University Press).

Bird, G. (1966), 'The Beginning of the Universe', *Proceedings of the Aristotelian Society*, Supplementary Volume XL, pp. 139–50.

Black, M. (1962), *Models and Metaphors* (Ithaca: Cornell University Press).

Blackburn, S. (1973), *Reason and Prediction* (Cambridge University Press).

Blokhintsev, D. I. (1972), *Space and Time in the Microworld* (Dordrecht: D. Reidel).

Bohm, D., Hiley, B. J., and Stuart, A. E. G. (1970), 'On a New Mode of Description Physics', *International Journal of Physics*, 3, pp. 171–83.

Boltzmann, L. (1964), *Lectures of Gas Theory*, trans. S. G. Brush (Berkeley: University of California Press).

Boscovitch, R. J. (1966), *A Theory of Natural Philosophy*, trans. J. M. Child (Cambridge: Massachusetts Institute of Technology).

Bouwsma, O. K. (1967), *Philosophical Papers* (Lincoln: University of Nebraska Press).

Boyd, R. (1973), 'Realism, Underdetermination, and a Causal Theory of Reference', *Nous*, pp. 1–12.

Boyd, R. (1976), 'Approximate Truth and Natural Necessity', *Journal of Philosophy*, LXXII, pp. 633–5.

Bradley, F. H. (1930), *Appearance and Reality* (Oxford University Press).

254

Brans, C., and R. H. Dicke (1961), 'Mach's Principle and a Relativistic Theory of Gravitation', *Physical Review*, CXXIV.

Bunge, M. (1967), *Foundations of Physics* (Heidelberg: Springer Verlag).

Burtt, E. A. (1924), *The Metaphysical Foundations of Modern Physical Science* (London: Routledge & Kegan Paul).

Čapek, M. (1960), 'The Theory of Eternal Recurrence in Modern Philosophy of Science, With Special Reference to S. C. Peirce', *Journal of Philosophy*, 57, pp. 289–96.

Čapek, M. (1961), *The Philosophical Impact of Contemporary Physics* (Princeton: van Nostrand).

Čapek, M. (ed.) (1975), *Concepts of Space and Time: Their Structure and their Development* (Dordrecht: D. Reidel).

Carnap, R. (1956), *Meaning and Necessity* (University of Chicago Press).

Carroll, L. (1960), *Alice's Adventures in Wonderland and Through the Looking Glass* (New York: New American Library).

Danto, A. (1965), *Nietzsche as Philosopher* (New York: Macmillan).

Dicke, R. H., and C. Brans (1961), 'Mach's Principle and a Relativistic Theory of Gravitation', *Physical Review*, CXXIV.

Dingle, H. (1964), 'Reason and Experiment in Relation to the Special Theory of Relativity', *British Journal Philosophy of Science*, 15, pp. 41–88.

Domoter, Z. (1972), 'Causal Models and Space-Time Geometries', *Synthese*, 24, pp. 5–27.

Dummett, M. A. E. (1954), 'Can an Effect Precede its Cause?', *Proceedings of the Aristotelian Society*, Supplementary Volume XXVIII, pp. 27–44.

Dummett, M. A. E. (1960), 'A Defence of McTaggart's Proof of the Unreality of Time', *Philosophical Review*, LXIX, pp. 497–504.

Dummett, M. A. E. (1964), 'Bringing about the Past', *Philosophical Review*, 73, pp. 338–59.

Dummett, M. A. E. (1969), 'The Reality of the Past', *Proceedings of the Aristotelian Society*, 13, pp. 239–58.

Dummett, M. A. E. (1973), *Frege: Philosophy of Language* (London: Duckworth).

Dummett, M. A. E. (1978), *Truth and other Enigmas* (London: Duckworth).

Earman, J. (1970), 'Space-Time, or how to solve Philosophical Problems without Really Trying', *Journal of Philosophy*, 67, pp. 259–77.

Earman, J. (1972), 'Notes on the Causal Theory of Time', *Synthese*, 24, pp. 74–6.

Ellis, B. (1955), 'Has the Universe a Beginning in Time?', *Australasian Journal of Philosophy*, 33, pp. 33–7 and 121–3.

Ellis, B. (1968), *Basic Concepts of Measurement* (Cambridge University Press).

Ellis, B., and Bowman, P. (1967), 'Synchronisation by Slow Clock Transportation', *Philosophy of Science*, 34, pp. 116–36.

English, J. (1973), 'Underdetermination: Craig and Ramsey', *Journal of Philosophy*, 72, pp. 343–63.

Bibliography

Eudemus (1870), *Eudemi Rhodii Peripatetici*, ed. L. Spengel (Berlin: Apud S. Calvary Eiusque Socium).

Findlay, J. N. (1941), 'Time: A Treatment of Some Puzzles', *Australasian Journal of Philosophy*, 19.

Føllesdal, D. (1973), 'Indeterminacy of Translation and Underdetermination of the Theory of Nature', *Dialectica*, 27, pp. 289–301.

Frege, G. (1950), *The Foundations of Arithmetic*, trans. J. L. Austin (New York: Harper).

Frege, G. (1971), 'On the Law of Inertia', *Studies in the History and Philosophy of Science*, 2, pp. 195–212.

Gale, R. M. (ed.) (1967), *The Philosophy of Time* (New York: Anchor Books).

Gale, R. M. (1968), *The Language of Time* (London: Routledge & Kegan Paul).

Gamov, G. (1954), 'Modern Cosmology', *Scientific American*, 190.

Glymour, C. (1972), 'Topology, Cosmology and Convention', *Synthese*, 24, pp. 195–218.

Glymour, C. (1976), 'To Save the Noumena', *Journal of Philosophy*, LXXIII, pp. 635–7.

Gödel, K. (1949), 'An Example of a New Type of Cosmological Solution of Einstein's Field Equations of Gravitation', *Review of Modern Physics*, 21, pp. 446–50.

Gold, T. (ed.) (1967), *The Nature of Time* (Ithaca: Cornell University Press).

Graves, J. C. (1971), *The Conceptual Foundations of Contemporary Relativity Theory* (Cambridge: Massachusetts Institute of Technology).

Grünbaum, A. (1967), *Modern Science and Zeno's Paradoxes* (Middleton: Wesleyan University).

Grünbaum, A. (1968), *Geometry and Chronometry in a Philosphical Perspective* (Minneapolis: University of Minnesota Press).

Grünbaum, A. (1973a), *Philosophical Problems of Space and Time*, 2nd Enlarged Edn (Dordrecht: D. Reidel).

Grünbaum, A. (1973b), 'Geometrodynamics and Ontology', *Journal of Philosophy*, LXX, pp. 775–800.

Grünbaum, A., *et al.* (1969), 'A Panel Discussion of Simultaneity by Slow Clock Transport in the Special and General Theories of Relativity', *Philosophy of Science*, 36.

Hamblin, C. (1971), 'Instants and Intervals', *Studium Generale*, 24, pp. 127–34.

Harré, R. (1962), 'Philosophical Aspects of Cosmology', *British Journal for the Philosophy of Science*, XIII, pp. 104–19.

Harré, R. (1970), *The Principles of Scientific Thinking* (London: Macmillan).

Hawking, S. W., and Ellis, G. F. R. (1973), *The Large Scale Structure of the Universe* (Cambridge University Press).

Hesse, M. (1974), *The Structure of Scientific Inference* (London: Macmillan).

Bibliography

Hill, E. L. (1955), 'Relativistic Theory of Discrete Momentum Space and Discrete Space-Time', *Physical Review*, 100, pp. 1780–3.

Hinckfuss, I. (1975), *The Existence of Space and Time* (Oxford: Clarendon Press).

Hollis, M. (1967), 'Time and Spaces', *Mind*, LXXVI, pp. 524–36.

Hooker, C. A. (1971), 'The Relational Doctrines of Space and Time', *British Journal for the Philosophy of Science*, 22, pp. 97–130.

Hume, D. (1960), *A Treatise of Human Nature*, ed. L. A. Selby-Bigge (Oxford University Press).

Hunter, G. (1971), *An Introduction to the Metatheory of Standard First-Order Logic* (London: Macmillan).

Kant, I. (1968), *Kant: Selected Pre-Critical Writings*, trans. G. B. Kerferd and D. E. Walford (Manchester University Press).

Kant, I. (1963), *The Critique of Pure Reason*, trans. N. Kemp Smith (London: Macmillan).

Kelly, J. L. (1955), *General Topology* (Princeton: Van Nostrand).

Kirk, R. (1972), 'Underdetermination of Theory and Indeterminacy of Translation', *Analysis*, pp. 195–201.

Lacey, H. M. (1968), 'The Causal Theory of Time, A Critique of Grünbaum's Version', *Philosophy of Science*, 35, pp. 322–54.

Leibniz, G. W. (1916), *New Essays Concerning Human Understanding*, trans. A. G. Langley (La Salle: Open Court).

Leibniz, G. W. (1956), *Leibniz-Clarke Correspondence*, trans. H. G. Alexander (Manchester University Press).

Lewis, D. (1976), 'The Paradoxes of Time Travel', *American Philosophical Quarterly*, 13, pp. 145–52.

Locke, J. (1961), *An Essay Concerning Human Understanding*, ed. J. W. Yolton (London: J. M. Dent & Sons).

Lucas, J. R. (1969), 'Euclides ab omni naevo vindicatus', *British Journal for the Philosophy of Science*, 20, pp. 1–11.

Lucas, J. R. (1973), *A Treatise on Time and Space* (London: Methuen).

Mach, E. (1960), *The Science of Mechanics*, trans. J. J. McCormack (La Salle: Open Court).

Mackie, J. L. (1955), 'Has the Universe a Beginning in Time?', *Australasian Journal of Philosophy*, 33, pp. 118–21, 123–4.

Mackie, J. L. (1974), *The Cement of the Universe* (Oxford University Press).

McTaggart, J. M. E. (1927), *The Nature of Existence* (Cambridge University Press).

Mann, T. (1928), *The Magic Mountain*, trans. H. T. Lowe-Porter (London: Secker & Warburg).

Manuel, F. E. (1965), *Shapes of Philosophical History* (London: Allen & Unwin).

Margenau, A. (1950), *The Nature of Physical Reality* (New York: McGraw-Hill).

Maxwell, J. C. (1953), *Matter and Motion* (New York: Dover).

Mendelson, B. (1968), *Introduction to Topology* (Boston: Allyn & Bacon).

Bibliography

Moore, G. E. (1963), *Some Problems of Philosophy* (London: Allen & Unwin).

Munitz, M. K. (1957), *Space, Time and Creation* (Glencoe: The Free Press).

Munitz, M. K. (1962), 'The Logic of Cosmology', *British Journal for the Philosophy of Science*, XIII, pp. 34–50.

Munitz, M. K. (1965), *The Mystery of Existence* (New York: Appleton-Century-Crofts).

Nerlich, G. (1976), *The Shape of Space* (Cambridge University Press).

Newton, I. (1953), *Newton's Philosophy of Nature*, ed. H. S. Thayer (New York: Hafner).

Newton-Smith, W. (1978), 'A Tense Logic for Closed Time' (unpublished paper).

Newton-Smith, W. (1979), 'A Tense Logic for the Special Theory of Relativity' (unpublished paper).

Nietzsche, F. (1968), *The Will to Power*, trans. W. Kaufman and R. J. Hollingdale (New York: Random House).

North, J. (1965), *The Measure of the Universe* (Oxford University Press).

North, J. (1970), 'The Time Coordinate in Einstein's Restricted Theory of Relativity', *Studium Generale*, 23, pp. 203–23.

Parfit, D. (1971), 'Personal Identity', *Philosophical Review*, LXXX, pp. 3–27.

Peirce, C. S. (1935), *Collected Papers of C. S. Peirce*, ed. C. Hartsthorne and P. Weiss (Cambridge: Harvard University Press).

Pike, N. (1970), *God and Timelessness* (London: Routledge & Kegan Paul).

Poincaré, H. (1913), *Dernières Pensées* (Paris: Ernest Flammarion).

Poincaré, H. (1946), *The Foundations of Science*, trans. G. B. Halstead (Lancaster: The Science Press).

Popper, K. R. (1956), 'The Arrow of Time', *Nature* 177, p. 538; *Nature* 178 (1956), p. 381; *Nature* 179 (1957), p. 1296; *Nature* 181 (1958), p. 402.

Prior, A. N. (1967), *Past, Present and Future* (Oxford: Clarendon Press).

Prior, A. N. (1968), *Papers on Time and Tense* (Oxford University Press).

Prior, A. N. (1970), 'The Notion of the Present', *Studium Generale*, 23, pp. 245–8.

Putnam, H. (1975a), *Mathematics, Matter and Method* (Cambridge University Press).

Putnam, H. (1975b), *Mind, Language and Reality* (Cambridge University Press).

Putnam, H. (1975–6), 'What is "Realism"?', *Proceedings of the Aristotelian Society*, 76.

Quine, W. O. (1960), *World and Object* (Cambridge: Massachusetts Institute of Technology).

Quine, W. O. (1966), *The Ways of Paradox* (New York: Random House).

Quine, W. O. (1969), *Ontological Relativity* (New York: Columbia University Press).

Quine, W. O. (1970), 'On the Reasons for Indeterminacy of Translation', *Journal of Philosophy*, pp. 178–83.

Quinton, A. (1962), 'Spaces and Times', *Philosophy*, 37, pp. 130–47.

Quinton, A. (1973), *The Nature of Things* (London: Routledge & Kegan Paul).

Reichenbach, H. (1947), *Elements of Symbolic Logic* (New York: Macmillan).

Reichenbach, H. (1950), *The Philosophy of Space and Time* (New York: Dover Publications).

Reichenbach, H. (1956), *The Direction of Time* (Berkeley: University of California Press).

Reichenbach, H. (1959), *The Rise of Scientific Philosophy* (Berkeley: University of California Press).

Rescher, N., and Garson, J. (1971), *Temporal Logic* (New York: Springer Verlag).

Rooteslaar, B. van, and Staal, J. F. (ed.) (1968), *Logic, Methodology and Philosophy of Science III* (Amsterdam: North Holland).

Royden, H. L. (1964), *Real Analysis* (London: Macmillan).

Rundle, B. (n.d.), 'The Past' (unpublished paper).

Russell, B. (1901), 'Is Position in Time and Space Absolute or Relative?', *Mind*, 10, pp. 293–317.

Russell, B. (1903), *The Principles of Mathematics* (London: Allen & Unwin).

Russell, B. (1914), *Our Knowledge of the External World* (London: Allen & Unwin).

Russell, B. (1936), 'On Order and Time', *Proceedings of the Cambridge Philosophical Society*, 32, pp. 216–28.

Russell, B. (1940), *An Inquiry into Meaning and Truth* (London: Allen & Unwin).

Russell, B. (1956), *Logic and Knowledge*, ed. R. C. Marsh (London: Allen & Unwin).

Schlegal, R. (1961), *Time and the Physical World* (New York: Dover Publications).

Scriven, M. (1954), 'The Age of the Universe', *British Journal for the Philosophy of Science*, 5, pp. 181–90.

Sellars, W. (1962), 'Time and the World Order', in *Minnesota Studies in the Philosophy of Science*, III (Minneapolis: University of Minnesota Press).

Shoemaker, S. (1969), 'Time without Change', *Journal of Philosophy*, LXVI, pp. 363–81.

Sklar, L. (1974), *Space, Time and Spacetime* (Berkeley: University of California Press).

Smart, J. J. C. (1963), *Philosophy and Scientific Realism* (London: Routledge & Kegan Paul).

Smart, J. J. C. (ed.) (1964), *Problems of Space and Time* (New York: Macmillan).

Smart, J. J. C. (1968), *Between Science and Philosophy* (New York: Random House).

259

Bibliography

Snyder, H. S. (1947), 'Quantized Space-Time', *Physical Review*, 71, pp. 38–41.

Stein, H. (1970), 'On the Paradoxical Time-Structures of Gödel', *Philosophy of Science*, 37, pp. 589–601.

Stoll, R. R. (1961a), *Sets, Logic and Axiomatic Theories* (San Francisco: W. H. Freeman).

Stoll, R. R. (1961b), *Set Theory and Logic* (San Francisco: W. H. Freeman).

Strawson, P. F. (1959), *Individuals* (London: Methuen).

Strawson, P. F. (1966), *The Bounds of Sense* (London: Methuen).

Swinburne, R. (1965), 'Time', *Analysis*, pp. 189–91.

Swinburne, R. (1966), 'The Beginning of the Universe', *Proceedings of the Aristotelian Society*, Supplementary Volume XL, pp. 125–38.

Swinburne, R. (1968), *Space and Time* (London: Macmillan).

Suppes, P. (1957), *Introduction to Logic* (New York: van Nostrand).

Suppes, P. (1972), 'Some Open Problems in the Philosophy of Space and Time', *Synthese*, 24, pp. 298–316.

Tarski, A. (1956), *Logic, Semantics, Metamathematics*, trans. J. H. Woodger (Oxford University Press).

Van Fraassen, B. C. (1970), *An Introduction to the Philosophy of Time and Space* (New York: Random House).

Van Fraassen, B. C. (1972), 'Earman on the Causal Theory of Time', *Synthese*, 24, pp. 87–95.

Van Fraassen, B. C. (1976), 'To Save the Phenomena', *Journal of Philosophy*, LXXIII, pp. 623–32.

Von Wright, G. H. (1969), *Time, Change and Contradiction* (Cambridge University Press).

Waismann, F. (1959), *Introduction to Mathematical Thinking* (New York: Harper & Row).

Weyl, H. (1949), *Philosophy of Mathematics and Natural Science* (Princeton University Press).

Wheeler, J. (1962), *Geometrodynamics* (New York: Academic Press).

Whitrow, G. J. (1961), *The Natural Philosophy of Time* (London: Thomas Nelson & Sons).

Whitrow, G. J. (1972), *What is Time?* (London: Thames & Hudson).

Wittgenstein, L. (1961), *Tractatus Logico – Philosophicus*, trans. D. F. Pears and B. F. McGuinness (London: Routledge & Kegan Paul).

Wittgenstein, L. (1975), *Philosophical Remarks*, trans. R. Hargreaves and R. White (Oxford: Basil Blackwell).

Zahar, E. G. (1973), 'Why did Einstein's Programme supersede Lorentz's? (I and II)', *British Journal for the Philosophy of Science*, 24, pp. 95–123 and 223–62.

Zdzislaw, A. (1967), 'Three Studies in the Philosophy of Space and Time', ed. R. S. Coehn and M. X. Wartofsky, *Boston Studies in the Philosophy of Science*, III (Dordrecht: D. Reidel).

Zeeman, E. C. (1964), 'Causality Implies the Lorentz Group', *Journal of Mathematical Physics*, 5, pp. 490–3.

INDEX

261